Our Future is Hydrogen!

ENERGY, ENVIRONMENT, AND ECONOMY

ROBERT SIBLERUD

New Science Publications
9435 Olsen Court
Wellington, CO 80549
(970) 568-7323

Our Future is Hydrogen!
Energy, Environment, and Economy

Printed and bound in the United States of America

Published by
New Science Publications
9435 Olsen Court
Wellington, CO 80549
(970) 568-7323

ISBN: 0-9666856-3-6

Library of Congress Catalog Card Number: 2001 130553

Acknowledgments:

Cover Art: **Ron Russell** (Copyright © 1998, Ron Russell)
Graphics: **Angela Phillips**
Editing: **Margaret Shaw, John LeCompte**
Copy Editing: **Betty Taylor**
Proof Editing: **Shirley Parrish**
Typesetting and Layout: **Pat Alles**
Dedication to the World Hydrogen Congress: **Heather Jarvis**
A True Visionary: **Maury Albertson, Ph.D.**
Hydrogen's Leader: **T. Nejat Veziroglu, Ph.D.**
Library: **Colorado State University**

TABLE OF CONTENTS

FOREWORD

T. Nejat Veziroglu, Ph.D.*

Fossil fuels (i.e., petroleum, natural gas and oil), which meet most of the world energy demand today, are being depleted fast. It is expected that within twenty years, production of petroleum and natural gas will peak, and then start to decrease. Although there are larger deposits of coal (to last a few hundred years), we may not be able to utilize them for environmental reasons.

Fossil fuels have another shortcoming. Their combustion products are causing global problems, such as the greenhouse effect, ozone layer depletion, acid rains and pollution, which are posing great harm to the environment, and eventually to life on our planet. The main combustion of fossil fuels, carbon dioxide emission, is causing climate changes, which in turn, results in natural disasters, such as more powerful and more frequent storms, floods, and droughts. These catastrophes are undermining the global economy. The insurance companies, which insure against such disasters, have been losing money since the 1980s. As a result, they are refusing to insure industrial plants, businesses, hotels, buildings, and homes in certain geographical areas, and/or increasing their premiums.

More than a quarter century ago, at the Hydrogen Economy Miami Energy (THEME) Conference of 18-22 March 1974, a handful of scientists formally proposed the Hydrogen Energy System as a solution to the interrelated global problems of energy and environment. Hydrogen is a very efficient and clean fuel. Its combustion will produce no greenhouse gases, no ozone layer depleting chemicals, little or no acid rain ingredients or pollution. Hydrogen, produced from renewable energy (direct and indirect solar) sources, would result in a permanent energy system, which we would never have to change.

In the intervening years, there has been some progress towards the Hydrogen Energy System. The space programs around the world use hydrogen as their fuel of choice for rocket engines. Metal hydride electric batteries for computers and cars have already been commercialized. Some types of fuel cells, which are not constrained by Carnot Cycle limitations, are already commercial; others are being developed. Major car companies have announced that in a couple years' time, they will market hydrogen fueled cars and buses. Hydrogen hydride air-conditioning systems are being developed, which will need no chloro-fluorocarbons or ozone layer depleting chemicals. Kitchen appliances based on catalytic combustion of hydrogen have already been developed. In other words, the conversion to the hydrogen Energy System has already begun.

Although the Hydrogen Energy System is the answer to global energy and environmental problems, and is expected to replace the fossil fuel system, the progress towards its realization is not too satisfactory. The reason for this is that the Hydrogen Energy System, its benefits, and its advantages are not widely known. Robert Siblerud, author of this excellent book, has clearly identified this fact by saying "Once the hydrogen story is known, the transition to our new (hydrogen) economy will progress rapidly." This book does just that, and it tells the hydrogen story in the best possible way. It gives the background information, the reasons for the hydrogen economy, hydrogen production, storage and distribution, safety, utilization in different energy sectors, available technologies, economics, and the timetable for the transition. It is clear, supported by facts, and written in a smooth and flowing style. It can be easily comprehended by the public-at-large. Once you read it, you will become a convert to hydrogen energy.

Our Future is Hydrogen! Energy, Environment, and Economy should be read by everyone concerned with the environment, quality of life, energy supplies, and the future of our home, the Planet Earth, the only one known to be hospitable to life. It will certainly help achieve its goal, viz., to speed up the transition to the hydrogen economy and a sustainable future.

*T. Nejat Veziroglu, Ph.D.
University of Miami
President, International Association for Hydrogen Energy
Nominee for 2000 Nobel Prize in Economics (for his work with hydrogen)

PREFACE

During the past decade, I have been fortunate to be involved, along with others, in a search for an alternative to fossil fuels. As director of the International Association for New Science, I have helped sponsor a number of new energy conferences, hoping to find a solution that would be environmentally and economically friendly. Surprisingly, we discovered many researchers who shared this vision and were also seeking the answer through new paradigm energy sources.

Over the years, our hopes were raised with the potential of cold fusion, zero-point energy, scalar energy, magnetic motors, electrostatic energy, Brown's gas, and many more. Many of these scientists were trying to find the "Holy Grail" of energy, over unity energy (getting more energy out of a system than is put in) that would be free to the public, much of it based on Nicholas Tesla's research. Many researchers made great strides on shoestring budgets, leaving one to imagine of potentially great discoveries with proper funding. A few scientists were fearful that once the breakthough was made, the technology would be suppressed. We heard many horror stories about promising patents being bought by big industry, only to be mothballed in fear driven by economics. Several researchers feared for their lives, and there were rumors of suspicious deaths. From our new energy conferences originated the Academy for New Energy and the Institute for New Energy, both organizations dedicated to providing a network for new energy scientists.

At several of our conferences, papers were presented on hydrogen energy. As we learned more about this alternative fuel, it became obvious it would offer the solution for our environmental dilemma. Hydrogen was a win-win situation for both the economy and the environment. Oil companies would be major players in the hydrogen economy, and most auto manufacturers have realized that hydrogen will be the fuel of the future.

For several decades, enlightened scientists have provided the foundation for our new hydrogen economy. Hydrogen technology has advanced greatly around the world through the efforts of organizations such as the International Association for Hydrogen Energy, the American Hydrogen Association, and the National Hydrogen Association. As we will see, both government and industry now realize hydrogen is the key to our future.

Once the hydrogen story is known, the transition to our new economy will progress rapidly. The one factor missing is public awareness of hydrogen. Now is the time to educate the public, the reason for this book. A news media blitz is planned for the next several years to increase public knowledge about hydrogen. In November 2001, a World Congress for a Hydrogen Economy will take place in Denver, Colorado, to help educate the public, government, academia, and industry.

The future for our planet is bright because hydrogen provides the solution to environmental pollution and dwindling fossil fuels. On the transition path from fossil fuels to hydrogen, the economy will prosper. The fossil fuel infrastructure will remain in place without loss of jobs, but a change only in job descriptions, as this infrastructure becomes the hydrogen infrastructure. Energy investors will prosper.

Thank you for letting us share with you the hydrogen story, and we are confident that you too will be inspired about the future of hydrogen.

Robert Siblerud

Chapter One

THE NEED FOR ALTERNATIVE ENERGY

We Have No Choice

"The sustainable production of hydrogen is the long-term pollution-free-energy- production future for America." Senator Tom Harkin of Iowa spoke these powerful words during hearings for the 1995 Hydrogen Futures Act. Hydrogen, our future energy fuel? Most people in America have never heard of using hydrogen as an energy fuel. Along with Senator Harkin and many of his congressional colleagues, leading scientists around the world conclude that hydrogen is the energy fuel of the future. Senator Harkin confirms this: "Many scientists believe that clean burning, efficient, and plentiful hydrogen is the energy carrier of the future." Two things necessitate the search for alternative fuels: (1) depletion of fossil fuel reserves, and (2) global warming. Humanity has undergone several energy evolutions in the past: from burning wood, to burning coal, to burning fuel oil, to burning natural gas, to using nuclear energy, and finally now to exploring sustainable nonpolluting, alternative fuels.

To comprehend the hydrogen story, we need to understand the harmful effects that burning fossil fuels have on the planet. We also need to examine available alternative fuels, including their benefits and pitfalls. Finally, after hearing the hydrogen story, we can then see how hydrogen will form the base of our future economy.

Fossil Fuels

IN THE BEGINNING

The hydrogen story began approximately 10 to 20 billion years ago with the Big Bang. From this enormous explosion originated all of the existing energy in the universe, including the Sun's energy. Hydrogen protons formed within three minutes after the Big Bang. Protons began to combine with electrons to form hydrogen atoms when the universe was about 300,000 years old. This process of combination continued until the universe was about 1 million years old. Astronomers believe the sun is about 4.6 billion years old and will keep shining for about another 7 billion years.

To understand the Sun's energy, we need to examine a nuclear reaction. Hydrogen and helium are the two principal constituents of the Sun. In the Sun, when the nuclei of hydrogen atoms closely approach one another, a nuclear reaction can take place, which means nuclei from different atoms combine with each other. Nuclear particles can rearrange themselves to form various nuclear end products. In a nuclear reaction, the internal energy of the original nuclear particles can be greater than the internal energies of the end products. According to the Law of Conservation of Energy, energy can neither be created nor destroyed. So if the energy of the end products is less than the energies of the original reactants, the reaction must have given off energy. When nuclear reactions produce energy, the energy given off is likely to be 1 million times greater than the energy generated by a chemical reaction, such as burning. A nuclear force is one million times stronger than an electrical force.

Radiant energy emitted from the Sun results from a series of nuclear reactions. During these fusion reactions, four protons from the nuclei of four hydrogen atoms convert into a helium nucleus consisting of two protons, two neutrons, and two positive electrons. The mass of the original four hydrogen protons is greater than the mass of the new helium nucleus and two positive electrons. This excess mass is converted into energy according to Einstein's equation, $E = mc^2$. This loss of mass is very small, but the conversion of one gram of hydrogen releases the same amount of energy as the combustion of twenty tons of coal.

Converting hydrogen into helium makes the major contribution to the vast amount of energy generated by the Sun. Gases in the Sun's core can reach 28 million degrees Fahrenheit. At this temperature, atoms collide with one another at great speeds, enabling them to fuse together. Once fusion begins, it heats the star, forming heavier elements in addition to helium. Each second, 5 million tons of the Sun's mass is converted into 30 million joules of energy.

Hydrogen can also exist in interstellar space as atoms of hydrogen gas and hydrogen molecules; they spread out at one atom or molecule per cubic centimeter.

THE SUN'S ENERGY

The source of energy that makes life possible on Earth is the 0.02 percent of incoming solar radiation that is absorbed by plant leaves. Once absorbed, photosynthesis converts this energy into carbon dioxide and water. This photosynthesis process forms the basic foods needed to sustain all forms of animal life. In other words, it forms carbohydrates from carbon dioxide and water found in the chlorophyll-containing tissues of plants exposed to light. Oxygen is an end product of photosynthesis.

The light absorbed by the leaves drives a chemical reaction that creates a series of other chemical reactions to eventually produce adenosine triphosphate (ATP). This molecule is the universal source of energy in all living things. Plants use ATP for metabolism, and with the assistance of chlorophyll, combine carbon dioxide and water to create sugars and starches needed for life by all living things. Less than 1 percent of the incoming light energy shining on plants goes into the production of food. Plants use only a small range of light wavelengths to stimulate reactions that produce starches and sugars.

Photosynthesis provides us with our food, and it also removes carbon dioxide from the atmosphere and replaces it with oxygen. Energy coming into a system cannot be destroyed and no additional energy can be created. The sun energy stored in food is measured by calories. The temperature of a body is known to be the measure of the average speed of the atoms and molecules of which the body is composed. A calorie is the unit of thermal energy required to raise the temperature of one gram of water by one degree centigrade. In a typical day, an adult needs approximately 2000 calories to sustain life processes. This energy is equivalent to burning a 100-watt light bulb for twenty-four hours.

Photosynthesis also has provided us with the raw materials for fossil fuels. This 0.02 percent of energy with 1 percent efficiency has created 90 percent of the energy we use today. Fossil fuels are the remains of organic matter that are millions of years old and have undergone substantial physical changes. They are storehouses of the sun's energy entrapped in organic matter. Fossil fuels include coal, oil, and natural gas, which provide the United States with 90 percent of its energy.

COAL

Coal was formed from vegetation, much of it woody, that fell to the ground. If these plants had not been protected somehow from interacting with oxygen, they would have eventually decayed into their original constituents of carbon dioxide and water. Scientists presume that water shielded some fallen vegetation from the atmosphere, either because plants grew close to bodies of water or because sudden flooding killed the vegetation and covered it. This began the process of coal formation.

Peat is a precursor of coal and is compacted vegetation that forms at the edges of deep lakes or in swamps when vegetation falls into the water, protecting it from oxygen. Often used as a fuel, peat begins coal formation.

Coal is created from peat or from other protected vegetable matter through the action of bacteria or of increased pressure resulting from Earth's upheaval. Coal progresses from lignite (brown coal) to bituminous and finally to anthracite, the coal of highest quality. Coal quality is determined by its carbon quality, with anthracite coal being between 90 to 98 percent pure carbon mixed with hydrogen. Coal is a hydrocarbon with a high carbon content. Some coal deposits are estimated to be 400 million years old.

Coal is the most plentiful of all fossil fuels and is also the most damaging to the environment when burned. Coal reserves are estimated by doubling the amount of known reserves and adding some factor for future discoveries. At the present rate of coal use, coal reserves will last several hundred years. If coal were the sole source of energy, it would probably last about 100 years. Estimates of reserves are difficult to make and depend on conservation practices, mining technology, price, and environmental regulations.

When coal is heated to a moderate temperature in the presence of steam, it can be converted to a combustible mixture of carbon monoxide, hydrogen, and methane. Coal can also be converted to a liquid gas (methanol) that can be substituted for gasoline by heating it to a high temperature with hydrogen, carbon monoxide, and a suitable catalyst. This liquid is very polluting and much more expensive than gasoline. During World War II, Germany did not have access to enough oil and used this method extensively to power its war machine.

Sulfur is the most serious pollutant from burning coal. Scrubbers in coal power plants remove part of the sulfur from the coal. Scrubbers are limestone-lined smokestacks that attract sulfur and absorb most of the sulfur pollutant.

One environmental problem of burning coal that cannot be overcome is that coal releases carbon dioxide when it is burned. Carbon dioxide is the major culprit in global warming. Another problem is, no matter how effectively coal is burned, it leaves ash and fine particulates, which are blown into the atmosphere.

During the twentieth century, 100,000 miners were killed in mining accidents such as cave-ins or explosions. This tragic figure does not include all the suffering and deaths caused by Black Lung disease among miners.

OIL

Oil (petroleum) has a history similar to coal, but its composition did not originate from woody plants. It comes from krill (planktonic crustaceans), diatoms, algae, and plankton. These microorganisms were shielded from oxygen in the atmosphere, initially by rock and limestone. Before oil formation, this organic matter was trapped in shale that formed from the consolidation of clay, mud, or silt. Petroleum deposits have had a great deal of oxygen removed from their compounds, which is responsible for their efficiency as an energy source. Bacteria, it is presumed, is responsible for removing the oxygen. With further chemical processing and increased pressure and bacterial action, the organic matter within shale became oil. Following its formation, the oil seeped into highly porous rocks, where it is found today. Many oil deposits have become capped by geological domes of impervious rock created by Earth movements. We still do not fully understand all of the mechanisms that formed these large underground deposits.

Oil exists interstitially among the pores of rocks and is analogous to liquid in a sponge. Oil, usually under pressure, flows into a well when it is drilled. Primary production occurs when sufficient oil pressure exists for the oil to remove itself spontaneously. When the pressure is too low, oil can be obtained by pumping the oil to the surface. It is also possible to force the oil to the surface by pumping water or gas into the reservoir to increase the pressure. This process is called secondary production. Primary production removes about 25 percent of the oil from a well, and secondary production increases the removal to 32 percent.

Exceeding all the known world oil reserves is oil shale. Colorado, Utah, and Wyoming have enormous oil shale deposits. Extracting oil from shale is extremely costly, making it economically unfeasible at the present time for a private company to tap these reserves. Shale contains oil known as kerogen, a substance tightly bound to the rock. Only after heating the oil to 400 degrees centigrade can the oil be drawn off. Heating the rock causes a "popcorn effect," which causes a greater volume of residue than the original rock.

Today, the United States imports more than one-half of its oil, with approximately half of that coming from the Middle East. In fact, one-fourth of the world's oil reserves are found in Middle Eastern countries. The consistent flow of oil is so important to the U.S. economy that the Department of Defense spends billions of dollars to ensure an uninterrupted flow. The United

States is the world's largest energy consumer and the second largest producer. It is estimated that by the year 2010, imports will increase to more than 60 percent. At that time, it is projected that 12.8 millions barrels of oil per day will be imported, compared with 6.7 million in 1990.

Oil reserves are estimated by doubling the amount of known oil reserves to allow for new sources to be discovered. Again, as with coal, it is difficult to estimate how long these petroleum reserves will last, and estimates vary. In 1992, proven U.S. oil reserves were estimated to last for only ten years, and world oil reserves were estimated to last for barely forty years. By 2020, if present trends continue, nearly 70 percent of world oil will be pumped from the Middle East. In 2000, Dr. T. Nejat Veziroglu, professor at Miami University, stated that known reserves of oil and natural gas would last for forty years at current consumption rates. If the exponential population growth and demand growth are taken into account, these reserves will last for only twenty-five years. Additional undiscovered reserves may add another thirty years. In any year, the amount of newly discovered oil is less than the increased amount used in that year over the previous year.

Of all the fossil fuels, oil yields the most energy per unit volume. Oil is the most easily transported, and the price of oil is very low for the amount of energy delivered. It is easy to understand why oil is the premier fossil fuel. Petroleum is the rootstock for many chemical processes, including production of plastics and medicines. For these uses, it is far more valuable than using it as a fuel. Many believe the prudent strategy would be to conserve oil reserves for nonenergy uses that could be developed into socially useful consumer products.

Many pollutants occur from burning petroleum products for transportation and power. These pollutants are responsible for smog, ozone changes, nitrogen oxide that causes acid rain, excess carbon monoxide, and carbon dioxide. Oil spills have caused many billions of dollars of damage to the environment. The most devastating result of burning fossil fuels is global warming, a real problem that is affecting the entire world.

NATURAL GAS

Natural gas, another fossil fuel, consists principally of methane, which is rich in hydrogen and poor in carbon. Methane has one carbon atom surrounded by four atoms of hydrogen. This makes natural gas an excellent fuel because it releases the smallest amount of carbon dioxide per unit energy of any of the fossil fuels. It releases 10 percent less carbon dioxide than gasoline. Natural gas contributes 30 percent less to the greenhouse effect than does oil, and 43 percent less than coal.

Natural gas is one of the end products in the chemical and physical processes that have produced oil and coal. Some methane pockets have no coal or oil near them, making their origin somewhat of a mystery. Some scientists postulate that methane is part of the primeval material of the universe, resulting in an almost infinite supply. Most scientists do not hold this view.

Many natural gas deposits exist deep within the earth. They are difficult and expensive to find, but inexpensive to exploit once discovered. Very little processing needs to be done to make natural gas suitable for use as a fuel. Water vapor, sulfur, and heavy hydrocarbons are removed from the gas before it is sent to its destination. At the present rate of consumption, known natural gas reserves are estimated to last about 150 years. Knowledge of untapped reserves is less accurate than oil reserve estimates, and natural gas reserves are quite widespread around the planet, many under the seabed.

Methane hydrate is another possible store of natural gas. These deposits are huge and widely distributed deep underground. Methane hydrate consists of single molecules of methane that are trapped within crystalline cages formed by frozen water molecules. A large deposit of methane hydrate has been discovered hundreds of feet below the seabed off the coast of North Carolina. Technology has yet to be developed to mine this natural resource and recover the methane.

One advantage of natural gas is its versatility. It could replace any fossil fuel application, including the powering of automobiles. However, storing an adequate amount of natural gas fuel in an automobile presents a major drawback. Should serious shortages of gasoline develop, natural gas could serve as backup for gasoline except that it lacks a distribution system similar to the gasoline stations. During the transition to alternative fuels, natural gas could be developed as a fuel for ground and airline transport.

During the 1940s, construction of natural gas pipelines began throughout the United States and continued into the 1960s. This provided an easy and inexpensive delivery system to many remote parts of the country.

Natural gas produces less carbon monoxide, fewer nitrogen oxides, fewer hydrocarbons, and less sulfur than its sisters, coal and oil. Despite its many virtues, natural gas is still a fossil fuel and contributes significantly to global warming.

Fossil fuels are a limited resource, and their reserves will eventually be depleted. As these resources become more limited, their cost will soar, mandating a transition to alternative fuels. At this time, fossil fuels are relatively inexpensive, making it quite difficult for an alternative fuel to penetrate the market. Powerful financial interests have been highly invested in fossil fuels, and until recently, they have been determined to maintain the status quo to protect their investment. As we will see, hydrogen presents a compromise.

During the transition from a fossil fuel economy to a hydrogen economy, fossil fuels will be an important source of hydrogen. The oil and gas companies will become energy companies. Oil investors will maintain their profits, and most importantly, the health of our planet will improve.

The Environment

GLOBAL WARMING

The question "Is global warming a reality or a farce fabricated by over-zealous environmentalists?" has been debated over the past several decades. Some deny it exists, some call for further studies, while the majority of scientists now claim global warming is upon us. Global warming is a direct result of our industrial revolution, powered by fossil fuels. Emissions from fossil fuels have increased greenhouse gases, such as carbon dioxide and nitrogen oxides, and have contributed to an increased temperature of the planet. Studies of global temperature have shown an increase of about 1 degree Fahrenheit since 1860. If no further action is taken, continued population growth, economic growth, and energy growth could result in global warming of 3 degrees to 8 degrees Fahrenheit by 2030 compared to pre-industrial temperatures. The ramifications are great, a catastrophe that will forever change our planet and lifestyle.

So what are the effects of this global warming? Speculated scenarios resulting from global warming include the following: (1) a change in precipitation patterns; (2) an increase in severe storms; (3) dislocation of suitable land for agriculture; (4) possible extinction of certain biological species and ecosystems; and (5) flooding of many coastal areas due to rising sea levels resulting from thermal expansion of the oceans, melting glaciers, and melting ice caps.

As the greenhouse gases accumulate and the Earth's average temperature increases, local climates will become more variable. Some areas will become warmer and drier, and some cooler and wetter. The middle and upper latitudes will see the greatest warming during winter. The number of uncomfortable summer days will also increase. In Chicago, an equivalent doubling of carbon dioxide could increase the average number of days over 90 degrees Fahrenheit from fifteen to fifty-six. Increased warming will produce greater precipitation in some areas such as Canada and Russia, which could receive more snow and rain. As we will see in later chapters, an economy based entirely on hydrogen energy could prevent global warming, but we first need to understand the principles underlying the greenhouse effect.

THE GREENHOUSE EFFECT

Solar energy, mostly in the form of visible light, is absorbed by the Earth's surface and reemitted into space as long-waved infrared radiation. Of the radiation intercepted by the Earth, 30 percent is reflected back into space, and 47 percent is absorbed directly by the Earth or by the greenhouse gases. These greenhouse gases reflect back to Earth the radiation originally reflected from the

Earth. The greenhouse effect is the ability of certain gases, primarily water vapor and carbon dioxide, to absorb the outgoing infrared radiation, which is translated to heat. As a result, this creates a higher atmospheric-equilibrium temperature than would occur in the absence of water vapor and carbon dioxide.

The average surface temperature of Earth is 17 degrees centigrade (62 degrees Fahrenheit) whereas the surrounding outer space is -270 degrees centigrade. When a warm body comes in contact with a cold one, the warm body is cooled. Without greenhouse gases such as carbon dioxide, methane, and water vapor, the Earth would be frozen. These gases return more than 84 percent of the energy, reflected outward by the Earth's surface, back to Earth from the atmosphere. These gases are transparent to the Sun's light, which passes through the atmosphere and reaches Earth. When reflected from Earth, the nature of the light changes so that it no longer can pass back through the greenhouse gases. This is because the greenhouse gases now absorb the reflected light and reradiate the light energy back to Earth. In other words, reflected light cannot escape through the greenhouse gases and remains trapped to warm the Earth. Continued increases in the amount of greenhouse gases in the atmosphere will have a disastrous effect on the planet.

Gases that have the ability to absorb and produce this effect are called greenhouse gases. Without these naturally occurring greenhouse gases, the Earth's temperatures would be -18 degrees centigrade (0 degrees Fahrenheit) instead of the present 17 degrees centigrade. Scientific evidence suggests that increasing concentrations of carbon dioxide and other trace greenhouse gases caused by human activity have enhanced the greenhouse effect. If the current exponential rate of carbon dioxide increases for an additional century, climate models predict that global temperatures may increase 18 degrees Fahrenheit as reported in the July 1993 journal *Nature*. In January 2001, a United Nations sponsored meeting, in Shanghai, China, reported global temperatures could rise as much as 10.5 degrees during the next century, dramatically higher than the 2.5 to 5.5 degrees previously thought five years ago. The U.N. report by the Intergovernmental Panel on Climatic Change (IPCC), consisting of scientists from ninety-nine countries, said "new evidence shows more clearly than ever that rising temperatures are the fault of industrial pollution, not changes in the sun or from other natural causes." The panel used computer models to make their predictions. According to Michael Oppenheimer, chief scientist at Environmental Defense, if temperatures reach the worst scenario as predicted by IPCC, they could remain there for a millennium. Vast ice sheets in Greenland and Antarctica could melt, raising the sea level more than 30 feet. Florida would be history and the eastern seaboard would be inundated.

Measurements of carbon dioxide levels since 1959 have shown that atmospheric carbon dioxide levels have been increasing at the rate of 0.3 percent per year. Today, carbon dioxide levels are increasing at a rate of 0.5 percent per year. The present concentration of carbon dioxide is about 350 ppm (parts per million), compared with the pre-industrial level of 285 ppm in 1800. The present level is the highest concentration in the past 160,000 years. By the middle of the twenty-first century, scientists estimate that carbon dioxide levels will approach 600 ppm, and within 70 years, the level of carbon dioxide will double. During the past two decades, researchers have discovered increasing concentrations of other greenhouse gases. These gases include methane (CH_4), nitrogen oxide, and a variety of chlorofluorocarbons (CFCs).

Carbon dioxide is the most abundant anthropogenic (manmade) greenhouse gas in the atmosphere. Other trace gases are much more efficient absorbers of infrared radiation than carbon dioxide. Scientists estimate that carbon dioxide contributes 49 percent to global warming, methane contributes 18 percent, nitrogen oxide contributes 6 percent, and chlorofluorocarbons contribute 14 percent. Even though the latter three comprise only 38 percent of the total atmosphere, they can trap heat a thousand times more efficiently than can carbon dioxide. These trace gases are increasing faster than carbon dioxide and caused about one-half of the additional greenhouse effect in the 1990s. Sources for methane include cattle production, decomposition of organic wastes, and the burning of fossil fuels and forests. Nitrogen oxides result from combustion of fossil fuels, and CFCs are used as coolants in refrigeration and air-conditioning, as insulating gas in foam insulation, and as propellants in aerosol spray cans.

Statisticians have found a strong correlation between global temperature and carbon dioxide changes. At this time, it is not clear which is the cause and which is the effect. Data from ice cores and ocean sediments show a significant correlation between carbon dioxide and global temperature. Some studies have shown that most of the temperature increases occurred prior to the big increase in carbon dioxide. Before 1990, the five warmest years in recorded history occurred in the 1980s. The decade of the 1990s was the warmest on record, with 1995 setting a record. Every month in 1998, the el Niño year, set a new temperature record, records that had been set during the previous decade. The la Niña year of 1999 was also a warm year, setting many heat records around the world. Between 1982 and 1988, ocean temperatures increased about 0.1 degree centigrade per year, a trend that continued in the 1990s, providing additional evidence that global warming is under way.

Some scientists claim that carbon dioxide entering the atmosphere today will be there for a century. Even if we stabilize carbon dioxide emissions now, the concentrations will continue to go up for a hundred years, along with increasing temperatures. Other scientists believe that eliminating carbon dioxide emissions from fossil fuels by replacing them with solar/hydrogen will reverse global warming.

A 1985 U.S. Environmental Protection Agency study suggested that global warming would cause forest systems to decline within a few decades. A warming of only 1.8 degrees Fahrenheit could shift some forest zones by more than 125 miles. In North America, a doubling of carbon dioxide could shift the range of birch, sugar maple, hemlock, and beech trees between 300 and 600 miles. This could threaten entire forests across the southern United States, making them more susceptible to disease and insects.

The consequence that global warming will have on the permafrost is often ignored. The melting of the permafrost could bring about the release into the atmosphere of 450 million tons of carbon in the form of carbon dioxide and methane. As a result, temperatures would rise and exacerbate global warming.

Gunter Weller, a retired University of Alaska geophysics professor and executive director of Arctic Climate Impact, is quoted by Associated Press in March of 2001 as saying that global warming has arrived in Alaska. Arctic winters have warmed much more quickly, as much as 10 degrees in the past 30 years, Weller said. "It is warmer, there is less sea ice. Permafrost is thawing. Glaciers are melting. There is less river and lake ice. The whole physical environment has changed very dramatically." Winter ice in the Bering Sea, Chukchi Sea, and Arctic Ocean is about half as thick as it was 30 years ago, claims Weller. The start of this year's Iditarod Dog Sled Race had to be moved north because warm weather made the trail too dangerous for the mushers and their dogs.

A number of species of plants and animals would die out. Animals could not migrate quickly enough to survive. Increased summer dryness in America's breadbasket would affect agriculture output. With a higher frequency of droughts and heat waves, yields of corn and wheat would decline. Water for drinking, irrigation, and industry would become more scarce. Nebraska farmers could abandon their fields for lack of water. Adjusting the world's irrigation systems alone would cost about $200 billion. By 2015, 3 billion people will be living in areas without enough water. Global warming will favor warm-weather species that have short life spans and evolve quickly. Weedy plants would thrive in a carbon-rich atmosphere, crowding out food crops. Warmer weather will cause insects, parasites, and pathogens to flourish, causing increased crop losses and more animal diseases. Malaria could be a public health threat in Vermont. Death from heatstroke worldwide may double by 2020.

Man-made catastrophes caused by global warming have increased from 70 per year in 1970 to between 225 and 230 per year today. The cost to insurance companies is enormous. In 1992, the damage from Hurricane Andrew cost $17 billion. An Associated Press release in January 2001 stated that in 1999, natural disasters in the form of mudslides, storms, and earthquakes took 105,000 lives. The disasters caused $100 billion worth of damage, of which $28.6 billion was insured. Most future catastrophes will be blamed on the burning of fossil fuels. In February 2001, the U.N. Environmental Program published a report stating that global warming could cost the world $304 billion annually in fifty years. In the 1990s, the cost was $608 billion, more than the four previous decades combined. The study claims that rising temperatures will trigger shifts in weather patterns causing natural disasters such as droughts, more frequent tropical cyclones, and rising sea levels.

Wild fires would be more common, similar to those in 2000 that burned 7 million acres in the western United States. Snow would melt earlier in the Rockies and Himalayas, exacerbating spring flooding.

Eight years of satellite data show a steady trend of ice-sheet shrinkage in the Antarctic. Scientists believe the ice cap atop Mount Kilimanjaro will disappear within fifteen years. Shrinkage has accelerated enormously during the past three years. In the Alps, scientists have estimated glaciers have lost 90 percent of the ice that was there a century ago.

Water levels of most of Africa's largest rivers, including the Nile, could plunge, and trigger widespread crop failure and idle hydroelectric plants.

If melting ice caps dilute the salt content of the sea, some scientists believe major ocean currents like the Gulf Stream could slow or even stop. This would inhibit the warming effect upon the Northern Hemisphere.

Most of the world's coastal wetlands will be lost to rising seas. Global warming is projected to raise sea levels between one and five feet, and by the year 2100, an Environmental Protection Agency study estimates the sea level will rise by seven feet. By mid century, hotels that grace Miami's South Beach could stand water logged and abandoned. Rising sea levels could contaminate the aquifer that supplies drinking water for the Caribbean nations. Entire Pacific island nations could disappear. Bangladesh would be the hardest hit, with 10 percent of the country becoming uninhabitable. The increase in global average temperature will be magnified two or three times in polar latitudes, accelerating the melting of ice caps. During the last century, the sea level has risen nine inches. The oceans rose six inches between 1955 and 1985, sometimes up to 0.5 inches per year. Today the oceans are rising at a faster rate.

Approximately 46 million people live at risk of flooding, due to storm surge. If the ocean rises twenty inches, that figure would double. The IPCC study predicts the seas will rise between 3.5 inches to 34.6 inches by 2010 because of thermal expansion (warmer water takes up more space) and because of melting glaciers and ice caps. A three foot rise would swamp parts of major cities and the Florida Keys. It would displace 4 million people in Nigeria. A relatively small rise in the sea level would contaminate groundwater with salt and would flood coastal roads, buildings, and wetlands. This would drive millions of people from their homes. Coastal communities would have to spend large amounts of money to protect against flooding. More than half of the Netherlands lies below sea level and would be threatened by storm surge. During the next decade, the Netherlands will spend more than $1 billion to build new dikes. Rising ocean surface temperatures would increase the frequency and severity of hurricanes and other tropical storms. The Great Barrier Reef could be ruined by a 3.6 degrees rise in water temperature.

Developing countries will be hit the hardest as flooding will dislocate millions. The United States, which contributes 25 percent to the greenhouse gases, will be affected less by climate changes than the tropical nations. Speculations and possible scenarios about the effects of global warming abound. Nothing is certain about what could happen, but it is certain that Earth is warmer today than before the Industrial Revolution.

FOSSIL FUEL POLLUTION

The greatest contributor to air pollution and anthropogenic gases is the burning of fossil fuels. Air pollution is any substance in the atmosphere at a high enough concentration to produce an objectionable effect on humans, animals, and vegetation, or to alter the natural balance of any ecosystem. In the early 1970s, concern over the chronic effects of air pollution led the United States to establish National Ambient Air Quality Standards (NAAQS) for six pollutants. These six pollutants include sulfur dioxide (SO_2), carbon monoxide (CO), nitrogen dioxide (NO_2), ozone (O_3), suspended particulates, and nonmethane hydrocarbons (NMHC), now referred to as volatile organic compounds (VOC). Evidence has linked high concentrations of these six pollutants to health problems. All six pollutants are by-products of fossil fuel combustion. Three of these pollutants, ozone, sulfur dioxide, and nitrogen dioxide, are phytotoxic, meaning toxic to vegetation. All six pollutants are easy to measure. Because nonmethane hydrocarbons are precursors to ozone, the Environmental Protection Agency (EPA) dropped them from the NAAQS list and added lead.

A framework for air quality management in the United States was established by the 1965 Clean Air Act (CAA), subsequently amended in 1970, 1977, and 1990. This act defined two categories of pollutants: (1) criteria pollutants and (2) hazardous air pollutants. For criteria pollutants, the Clean Air Act required the Environmental Protection Agency (EPA) to establish National Ambient Air Quality Standards (NAAQS) and emission standards for motor vehicles and large new sources of emissions. The EPA was responsible for designing and implementing air quality improvement. The NAAQS applied uniformly across the United States, but the emission standards for criteria pollutants varied, depending on the severity of local air pollution problems. The Clean Air Act declared that every five years, the scientific basis for NAAQS must be renewed. National, state, and local monitoring systems have been established by the EPA.

The twenty-four industrial nations that belong to the Organization of Economic Cooperation and Development accounted for about one-half of the following pollutants during 1980: 110 million tons of sulfur oxides, 69 million tons of nitrogen oxides, 193 million tons of carbon monoxide, 57 million tons of hydrocarbons, and 59 million tons of particulates. Burning fossil fuels around the world each year creates 30 billion tons of soot, ash, carbon dioxide, and other gases.

Sulfur Oxides

The combustion of sulfur-containing fossil fuels, especially coal, is the primary source of sulfur oxides. Natural sources include volcanoes and decaying organic matter. Between 97 and 99 percent of sulfur oxides emitted from combustion sources come in the form of sulfur dioxide (SO_2), one of the criteria pollutants. The remainder comes as SO_3, which is immediately transformed into sulfuric acid, H_2SO_4, a liquid particulate that plays a role in acid rain, to be discussed later.

Nitrogen Oxides

Nitrogen is a very common element. About 80 percent of the air we breath is composed of nitrogen. Nitrogen oxides are formed when combustion in air is greater than 4000 degrees Fahrenheit. Approximately half of the anthropogenic nitrogen oxides originate from motor vehicles, and one-third come from power plants. At the high temperatures created in the combustion process, some nitrogen (N_2) is converted to NO in the presence of oxygen. Most of the nitrogen oxides are emitted as NO and NO_2, and NO is oxidized to NO_2 in the atmosphere. Natural sources include lightning and decomposing organic matter.

NO_2 is one of the original criteria pollutants, a precursor to ozone. Nitrogen oxides are an essential ingredient of photochemical smog and are oxidized to nitrous acid (HNO_3), a factor in acid rain.

Carbon Monoxide

Carbon monoxide is a colorless, odorless gas emitted during the incomplete combustion of fuels. Transportation sources account for about two-thirds of non-natural carbon monoxide emissions, and other combustion sources such as wood burning, fireplaces, and stoves contribute another 25 percent. Carbon monoxide has little direct effect on ecosystems, but it contributes directly to the greenhouse effect and to depletion of the Earth's ozone layer. Between 60 and 90 percent of total carbon monoxide emissions come from natural sources.

Carbon monoxide is absorbed through the lungs into the bloodstream and reacts with hemoglobin to form carboxyhemoglobin, which reduces the oxygen carrying capacity of the blood. Health effects range from neural-behavioral effects, headaches, fatigue, respiratory failure, and death.

Volatile Organic Compounds

Volatile organic carbon compounds include any organic

compound that exists as a gas in air and can participate in photochemical reactions. Vehicles emit volatile organic compounds through evaporation of fuel and in the exhaust gases. These emissions represent unburned fuel or incomplete combustion of fossil fuels. Fuel combustion from power plants contributes 39 percent of VOCs, transportation contributes 33 percent, and industrial processes contribute 12 percent.

The primary risk associated with VOCs is that they react with oxides of nitrogen in the presence of sunlight to form photochemical oxidants such as ozone. These oxidants irritate mucous membranes, inflame the eyes and upper respiratory tract, cause chest constriction, and at high concentrations cause severe coughing. Some compounds can cause mutations, birth defects, and cancer.

Ozone

Hydrocarbons contribute to the formation of photochemical ozone, which is a principal component of smog. Ozone forms when volatile organic compounds (VOCs or hydrocarbons) and nitrogen oxides react with oxygen in the presence of sunlight. Ozone plays a role in acid rain because it contributes to the generation of radicals that convert oxides to nitric acid and sulfuric acid.

Ozone is a lung irritant and a phytotoxicant responsible for crop damage and suspected of contributing to forest decline in Europe and parts of the United States. Ground level ozone has been linked to the damage of 87 percent of ponderosa and Jeffrey pines in Southern California. It inhibits photosynthesis and increases plant susceptibility to disease. The U.S. Crop Loss Assessment Program estimates that damage to corn, wheat, soybeans, and peanuts caused by ozone results in losses of between $1.9 and $4.5 billion annually. Others put this figure at $5.6 billion. Ozone is an important greenhouse gas that causes eye irritation, nasal congestion, and lung disorders.

Ozone, although a hazard at ground level, acts as a filter in the upper atmosphere by screening out harmful radiation from the sun. In the stratosphere (nine to thirty miles above the earth's surface), ozone forms naturally when oxygen is dissociated by ultraviolet light in the wavelength range of 180 to 240 nanometers. The two free oxygen atoms that result from this reaction quickly bind to another oxygen to form ozone (O_3). This process is reversed when ultraviolet light breaks down ozone to form O_2 and O_1 creating a balance between O_1, O_2, and O_3. When other reactive substances such as chlorine and nitrogen oxide are present in the upper atmosphere, they upset the balance and reduce the amount of ozone.

The concentrations of ozone are only a few molecules to every million molecules in the stratosphere. Ozone is responsible for the removal of ultraviolet B radiation (wavelength = 280 to 330 nanometers) that could otherwise reach the earth's surface.

The first concern over ozone depletion resulting from nitrogen oxide emission occurred when a fleet of supersonic transport aircraft flew through the stratosphere and destroyed ozone. Nitrogen oxide acts as a catalyst and participates in the reaction sequence many times. In the mid-1970s, it was discovered that chlorofluorocarbons (CFCs) diffused unaltered through the troposphere into the mid-stratosphere, where they were photolyzed by ultraviolet radiation, resulting in chlorine and chloride/oxygen radicals. The chlorine serves as a catalyst and destroys ozone molecules. Each chlorine atom will destroy approximately 100,000 ozone molecules. CFC molecules are stable and can remain intact for more than 100 years to destroy the ozone layer. During the spring in recent years, researchers have observed ozone depletion of 60 percent at all altitudes over Antarctica, and depletion of 95 percent in some layers.

Chlorofluorocarbons are found in aerosol propellants, and in 1976, the United States banned their use. In 1990, ninety-three nations agreed to accelerate the phase out of CFCs and to completely eliminate their production by the year 2000.

Among the various types of ultraviolet (UV) radiation normally blocked by the ozone layer, UV-B is the most harmful and is known to cause skin cancer. The EPA estimates that during the next century, 800,000 U.S. cancer deaths and an increased number of cataracts may be attributed to depletion of the ozone levels and the resultant increase in UV-B radiation. Presently, UV-B annually kills an estimated 100,000 people worldwide. Each 1 percent drop in stratosphere ozone may increase the rate of skin cancer by 4 to 6 percent. By the year 2075, ozone depletion could cause 3 million to 15 million new cases of skin cancer.

In plants, UV exposure can decrease photosynthesis, stunt leaf growth, damage seed quality, and reduce crop yields. UV-B can kill large numbers of aquatic organisms, which form the base of the ocean's food chain. Many species of phytoplankton and fish larvae cannot tolerate an increase in UV-B.

Carbon Dioxide

Carbon dioxide is by far the most abundant anthropogenic greenhouse gas in the atmosphere. It is produced by fossil fuel combustion, and scientists estimate that it contributed to about half of the global warming in the 1980s. Internal combustion engines and combustion plants and furnaces are responsible for more than 95 percent of carbon dioxide and nitrogen oxide emissions.

In 1894, Swedish chemist Svante Arrhenius put forth the idea that carbon dioxide might serve as a thermostat for the atmosphere. Around 1950, the scientific community became concerned about the dangers of greenhouse warming caused by the increases in carbon dioxide. However, many scientists did not take this seriously. They believed water vapor, a greenhouse gas, occurs at a much higher concentration than does carbon dioxide in the atmosphere. They also thought that burning fossil fuels added little carbon dioxide to the greenhouse gases.

Scientists believed that because the ocean contained fifty times as much carbon dioxide as the atmosphere, it could serve as a reliable regulator of the atmospheric concentration. However, researchers later found that water vapor did not block the same radiation, so carbon dioxide from fossil fuels did have an important effect. Increased carbon dioxide levels in the atmosphere correlated with an increase in average temperature around the world. Studies have concluded that if we maintain our present rate of emissions, carbon dioxide concentrations will double by the middle of the twenty-first century, increasing the average temperature by 4.5 to 6.0 degrees Fahrenheit. At present, almost all scientists agree that global greenhouse warming is a reality, caused by increased concentrations of carbon dioxide. Ranga Mynen reported in *Nature* that between 1981 and 1991 an increase in the peak growing season occurred north of the 45th parallel. Spring was arriving earlier and fall was departing later.

Island nations have reported diminishing coastlines. During the 1980s, the insurance industry paid out less than $2 billion a year for weather-related damage. Between 1990 and 1995, they paid out $30 billion a year.

Smog

Photochemical smog is a complex mixture of constituents formed when volatile organic compounds (VOCs) and nitrogen oxides are irradiated by sunlight. From this effect, ozone presents the primary concern and is the most abundant pollutant found in photochemical smog. Motor vehicle exhaust is a major source of the nitrogen oxides and VOCs that create smog. According to a 2001 study, 141 million Americans are affected by smog and ozone, compared to a 1988 EPA study that found 110 million Americans breathed air with unsafe smog levels.

Photochemical smog is a summertime phenomenon in most parts of the United States, because during the other seasons, lower temperatures and insufficient sunlight exist. In the presence of sunlight, nitrogen oxide photolyzes and produces ozone. To reduce ground level ozone in a polluted atmosphere requires reducing VOCs and nitrogen oxides.

In 1991, the three areas with the highest levels of ozone were Los Angeles (0.31 ppm), Houston (0.22 ppm), and New York (0.17 ppm), with levels above 0.12 ppm per hour being detrimental to health. In the Los Angeles basin, high concentrations of ozone are transported throughout the basin, and multidaily episodes are exacerbated by the accumulation of ozone aloft, which can mix with the surface air each day. In 1990, the Clean Air Act amendment required that ozone non-attainment areas reduce both VOCs and nitrogen oxides from large stationary sources.

Particulates and Visibility

Particulates are finely divided solids and liquids dispersed in the air that contribute to poor visibility. The only common light-absorbing gaseous pollutant is nitrogen dioxide, which accounts for a few percentage points of total visibility extinction. In 1977, the Clean Air Act amendment set a goal to prevent future impairment of visibility.

Air pollutants can also cause discoloration of the atmosphere. The most common is a brownish discolorization found in the brown Los Angeles haze and the "brown cloud" over Denver. Several factors contribute to the brown color, including nitrogen oxide, a brownish gas. In urban hazes, the effect of nitrogen dioxide is usually overwhelmed by the effects caused by particles. Fine particles preferentially scatter blue light in the forward direction. The light observed through an optically thin cloud with the sun behind the observer is deficient in the blue wavelength, thus appearing brown.

Acid Rain

Acid rain has focused attention on ecosystem toxicology more than any other contaminant. Acid rain has a devastating effect on humans, animals, vegetation, and physical objects. Epidemiological studies, human studies, and animal studies all suggest that high exposure to acidic atmospheric suspensions may produce adverse health effects. High-risk populations include the elderly, the very young, and anyone with a chronic-lung dysfunction. Acid rain can result in significant deterioration of limestone, marble, bronze, unpainted galvanized steel, and numerous paints. Sulfates from burning fossil fuels contribute sulfuric acid to acid rain causing more than 50 percent of the light extinction in rural areas of the eastern United States.

At room temperature, pure water dissociates to produce equal amounts of hydrogen (H+) and hydroxyl ions (OH-). Acid solutions in water are defined as those solutions that have hydrogen ion concentrations greater than the hydroxyl concentrations (pH<7). When hydrogen and hydroxyl ion concentrations are equal, the acid-base balance pH = 7. A change from a pH of 7.0 to 6.0 represents a tenfold increase in hydrogen ion concentrations. In other words, it is acidic. Natural rain, including precipitation in relatively unpolluted regions, is naturally acidic, with a pH of 5.0 to 6.0. This natural acidity results from the oxidation of carbon oxides and subsequent formation of carbonic acid.

In regions downwind from electric generating power stations

that use fossil fuels, in industrial regions, or major urban centers, precipitation can be acidified below a pH of 5.0. Precipitation below 5.0 is designated as acid rain. The human-caused acidification of precipitation results from release of sulfur dioxides and nitrogen oxides from smokestacks and tailpipes. Sulfur and nitrogen oxides are oxidized to sulfate and nitrate and return to the earth as sulfuric and nitric acids. Acidification of precipitation by humans can reduce the pH of rain in downwind regions to between 4.0 and 5.0. The atmosphere deposits acidity onto land during and in between rain showers. The latter is termed dry deposition and the former is wet deposition.

Some forests are at special risk to the adverse effects of acid deposition. The forests receive high exposure to hydrogen, sulfate, and nitrate ions because of their proximity to primary sources or because of their higher elevations. High-elevation conifer forests along the Appalachian Mountain chain in the eastern United States are exposed to high levels of acid deposition because of frequent fog immersion and cloud-water deposition. The average acidity of cloud moisture is ten times greater than of moisture at lower elevations.

For most American forests, acid deposition influences on forest systems are neutral. Most surface water in the United States has sufficient buffering or acid neutralizing capacity to neutralize acidification by way of atmospheric deposition. Across the United States, the mean pH varies from 4.2 in western Pennsylvania to 5.7 in the West. The Adirondacks and mid-Atlantic highland regions have both sensitive aquatic resources and high levels of acid deposition. Ten percent of the lakes in the Adirondack Mountains have pH levels below 5.0. Thousands of lakes on the eastern seaboard have been labeled fish graveyards. At elevations greater than 2,900 feet, 20 percent of all spruce alive in 1982 had died by 1987 in the northeastern United States.

European nations have been hit especially hard by acid rain. In Sweden, fish have disappeared from 4,000 lakes, and another 20,000 lakes have been damaged by acid precipitation. Likewise in Norway, 80 percent of the lakes have been declared technically dead or placed on the critical list. Sweden claims 80 percent of its sulfur oxides come from other countries, and Norway claims 90 percent. Sulfur dioxide emissions can travel up to 1200 miles in a few days. In Switzerland, 43 percent of the conifer forests of the central alpine region are already dead. Approximately half of all trees in West Germany suffer from acid deposition. In Great Britain, two-thirds of its trees suffer from acid rain. In Poland, acid deposition is eroding iron railroads, forcing trains to travel at 25 mph. Acid rain is dissolving ancient monuments in Athens and Cologne.

Deposition of acid damages leaf surfaces and prevents some tree species from retaining water. Acid waters leach minerals such as calcium, magnesium, and potassium from leaves and the soil, depriving plants of vital nutrients. It also releases aluminum from the soil, which can damage tree roots, block nutrient absorption, and impair water transport. This makes trees more susceptible to drought, insects, and stress. Aluminum can filter into streams and lakes, contaminating fish and threatening public water supplies. Acidified water can kill many species of freshwater fingerlings and larvae. This disrupts the food chain.

In saltwater, nitrates from acid deposition boost the nitrogen content of coastal estuaries. This causes algae to bloom, which results in oxygen deprivation and suffocates fish and other aquatic plants.

Economic Costs

The economic costs resulting from fossil fuel pollution are great. Damage to metals, building facades, and paint costs the 24 nations that belong to the Organization of Economic Cooperation and Development an estimated $20 billion a year. The European community estimates that acidic deposition costs at least $1 billion in crop loss each year. Damage to timber and agriculture amounts to $1.4 billion per year. Canada estimates its acid rain damage amounts to $250 million annually, whereas the United States estimates damage to forest, agriculture, and aquatic ecosystems may exceed $5 billion each year.

In 1995, European countries agreed to reduce emissions of sulfur oxides by 60 percent and nitrogen oxides by 40 percent. An overall assessment by these countries concluded that the combined effects of airborne acids, ozone, and other air pollutants may reduce agriculture and forest productivity over large areas in industrial countries by 5 to 15 percent below historic levels.

Environmental factors dictate the necessity to replace fossil fuels with alternative fuels. The ideal alternative would be a fuel that is plentiful, produces no pollution, prevents global warming, and cleans the air. As we will see, hydrogen is that fuel.

Chapter Two

HYDROGEN

An Amazing Story

Some people are calling hydrogen the perfect fuel, and scientists are coming to the conclusion that hydrogen may be the savior energy of our planet. Hydrogen is plentiful, it is an inexhaustible resource, it is clean and non-polluting, and it is an excellent energy carrier. Hydrogen is not a primary source of energy, but is a secondary form. It becomes an energy carrier for all other energy sources, whether fossil, nuclear, or solar energy. Atoms of hydrogen are almost always found combined with other atoms (as in plants, in water, and in petroleum), and energy must be used to break the chemical bonds and produce pure hydrogen. This is why hydrogen is considered a storage medium.

When hydrogen is produced from some form of renewable energy based on the sun, such as solar thermal energy, wind energy, or hydroelectric energy, it is called "solar hydrogen." Wind power, solar power, photovoltaics, and hydroelectric power are all well suited to generate hydrogen from electrolysis of water. Solar hydrogen is a non-polluting, renewable energy carrier, and when burned, it produces only water and, in some instances, oxides of nitrogen, which are both natural in our atmosphere. Solar hydrogen energy is the most environmentally compatible energy system, and it may reverse global warming (if it replaces all fossil fuels) and bring the earth's temperature toward normal by decreasing carbon dioxide emissions. The technology is available, and the following chapters will provide a summary on making this transition to a hydrogen economy. But first, we need to understand the properties and history of hydrogen.

HYDROGEN PROPERTIES

Hydrogen is the most abundant element in the known universe. It is the simplest element in the periodical table and the smallest atom, consisting of one proton and one electron in the elemental state. The atomic weight of hydrogen is 1.0079, and its atomic number is 1. Atoms of an element that have a varying number of neutrons in their nuclei are called isotopes of that element. The isotopes of hydrogen are protium with no neutrons, deuterium with one neutron, and tritium with two neutrons, the latter being radioactive.

Earth's atmosphere contains only 0.00005 percent pure hydrogen, as hydrogen is normally bound to other atoms. Because hydrogen has such a small mass, it can escape the Earth's gravitational pull and fly off into space.

Henry Cavendish discovered hydrogen in 1766 by putting acid on iron and collecting the bubbles that it gave off. These bubbles proved to contain a gas (which was later identified as hydrogen) that not only burned but sometimes exploded in air. Because burning hydrogen produced water, Antoine Lavoisier called the element hydrogen, which in Greek means "water forming."

Hydrogen is a colorless, odorless, and tasteless chemical, usually a gas. It can be changed into liquid under much pressure and at a very low temperature, -423 degrees Fahrenheit, or -253 degrees centigrade. It freezes at -434 degrees Fahrenheit (-259 degrees centigrade).

Hydrogen comprises an essential part of all plant and animal bodies and most of our food. It is found in water and all acids. Very little hydrogen exists in the soil and rocks, and almost none in the air, except as part of water vapor. The Sun's atmosphere and other stars contain great quantities of hydrogen. The universe consists of 90 percent hydrogen atoms. Being the lightest known substance, hydrogen weighs 14.5 times less than air.

Pure hydrogen exists as hydrogen gas, in which pairs of hydrogen atoms are bound together to make molecules. Hydrogen can combine chemically with almost every other element and forms more compounds than any other element. It is found in water, minerals, organic matter, and hydrocarbons. It is the tenth most common element on earth, but accounts for less than 1 percent of the Earth's mass. Nearly every compound in the human body contains hydrogen.

Like most gaseous elements, hydrogen is diatomic, meaning its molecules contain two atoms. In a hydrogen molecule, the two nuclei share two electrons. The bond between the hydrogen atoms is very strong and breaking it requires a large amount of energy. When heated with a flame or a spark, hydrogen gas reacts vigorously with oxygen in the air to produce water. Hydrogen has the widest flammability limits of combustion in air. A volume of air that contains between 4 and 75 percent hydrogen can be ignited and support combustion.

When combusted, hydrogen offers emissions free of the major contributors to atmospheric pollution including unburned hydrocarbons, carbon monoxide, carbon dioxide, oxides of sulfur, and reduced amounts of nitrogen oxides. It does not contribute to acid rain, global warming, particulate emissions, or odor pollution.

The principal source of hydrogen is water, in which it is tightly bound to oxygen. To liberate hydrogen requires a great amount of energy.

The versatility of hydrogen makes it an excellent candidate for a universal fuel, for space exploration, for power stations, for powering vehicles, for industry, and for all fuel-based appliances. It is three times more efficient than gasoline. If the electricity to produce hydrogen from water is produced from solar energy, hydrogen in effect becomes a way of storing the power of the Sun, because the hydrogen produced can then be used when the Sun is not shining. Hydrogen can be used as a fuel for any energy or power-generating source for which we now use conventional fuels.

HYDROGEN FUEL PROPERTIES

Internal Combustion Engines

Before we continue, it might be wise to explain how an internal combustion engine (ICE) operates so we can lay the foundation for how these engines use hydrogen fuel. An ICE typically compresses the fuel and air in a cylinder to make a high-pressure mixture that is burned to create a hot, high-pressure gas, a product of combustion. As it burns, this gas expands to drive the piston and power the vehicle.

All present ICEs used in transportation use a slider-crank mechanism in which a piston slides along the axis of a cylinder turning the crank shaft to do work. Reciprocating ICEs can be classified as either a four-stroke cycle or a two-stroke cycle.

In the four-stroke cycle, the piston traverses the cylinder four times (up and down, two times) for each cycle the engine executes. This turns the crank shaft two revolutions. There are five steps:

1. With the *intake stroke,* the piston travels from "top dead center" to "bottom dead center" of the cylinder, and fuel air is drawn into the compression chamber of the cylinder past the intake valve.
2. In the *compression stroke,* the intake valve closes, and the fuel air charge is compressed by the piston moving up to "top dead center."
3. *Ignition,* in a spark ignition engine, is initiated by the occurrence of a spark between the electrodes of a spark plug that protrudes into the combustion chambers. In a compression ignition, the combustion process is initiated by injecting fuel into the compressed air.
4. During the *expansion stroke,* the combustion products expand as the piston moves down to the "bottom dead center." The exhaust valve opens, and the pressure in the cylinder drops as the gases rush out of the cylinder.
5. During the *exhaust stroke,* the piston moves toward the "top dead center," and as it does, it pushes the gases out of the

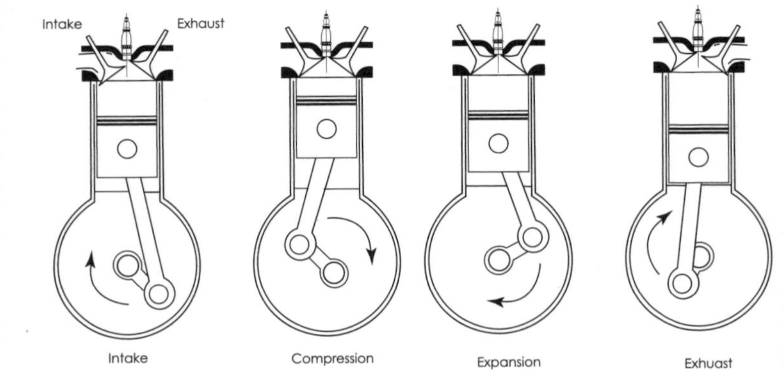

FOUR-STROKE CYCLE ENGINE

cylinder past the exhaust valve. As the piston approaches "top dead center," the intake valve starts to open and the exhaust valve begins to close.

Hydrogen Fuel

Limits of flammability are one of the most important properties of a fuel. These parameters measure the range of fuel/air ratios over which an engine can operate. Hydrogen has a wide range of flammabilty compared with other fuels. A significant advantage is that hydrogen can operate on a lean mixture. This lean mixture means the amount of fuel is less than the theoretical or chemically optimum amount needed for combustion with a given amount of air. In other words, it is an air-rich mixture. Greater fuel economy and a more complete combustion reaction results when an engine runs on a slightly lean mixture. The final combustion temperature can be lowered by using an ultra-lean mixture, which reduces the engine nitrogen-oxide emissions. One disadvantage is that a lean operation can significantly reduce the engine power output. However, hydrogen can be run more effectively on excessively lean mixtures than can gasoline engines. With hydrogen, as little as 4 percent in volume can produce a combustion mixture, compared with the 29.53 percent needed for stoichiometric operation with gasoline.

The minimum energy required for ignition is another important property of a fuel. For hydrogen, the minimum energy required for ignition is about an order of magnitude lower than that required for gasoline. This property allows a hydrogen engine to run well on a lean mixture and ensure prompt ignition. Because so little energy is necessary to ignite a hydrogen combustion reaction, almost any hydrogen/air mixture can be ignited because of hydrogen's wide limits of flammability. However, for some systems this can result in hot gases and hot spots on the cylinder wall that serve as a source of ignition, creating problems such as premature ignition and flashback.

In the cylinder (combustion chamber), the combustion flame is typically extinguished at a certain distance (called the "quenching distance") from the cylinder wall due to heat losses. The quenching distance for hydrogen is less than that of gasoline, so the flames come closer to the cylinder wall with hydrogen before it is extinguished. This smaller distance increases the tendency for backfire because the flame from a hydrogen/air mixture can more readily get past a nearly closed intake valve than can a flame from a hydrocarbon mixture.

The self-ignition temperature is the temperature that a combustion mixture reaches before it will be ignited without an external source of energy. This self-ignition temperature for hydrogen is relatively high, allowing larger compression ratios to be used in a hydrogen engine without increasing the final combustion temperature beyond the self-ignition temperature. As a result, this can cause premature ignition.

The flame speed for hydrogen is much greater than that of gasoline. Hydrogen engines can therefore approach the thermodynamically-ideal engine cycle. Diffusivity of hydrogen, its ability to disperse in air, is considerably greater than that of gasoline. These advantages facilitate the formation of a uniform mixture of fuel and air. If a leak develops, the hydrogen will disperse rapidly.

In summary, preignition occurs because hydrogen has a low minimum ignition temperature and a wide flammability range. Contact between the air mixture and hot residual gas causes an engine to backfire. Both of these dilemmas have presented major problems for the hydrogen engine. During the past several decades, research has solved these stumbling blocks. Hydrogen can now be added to the air mixture at a point where conditions for preignition are less favorable. Water injection or residual exhaust gas recirculation are added to hydrogen engines to control preignition.

Historically, four general fuel delivery systems have been used for hydrogen engines: (1) carburetors, (2) intake manifold injection, (3) intake port injection, and (4) direct cylinder injection. The first three methods involve forming the fuel-air mixture during the intake stroke, either through the carburetor, through the intake manifold, or through the intake port. However, by directly injecting fuel into the cylinder, fuel delivery can be controlled to take place after closure of the intake valve. Carburetion is no longer a viable technology for modern vehicles because of the need for very accurate mixture and burn timing for power and emission controls. It was very easy to convert a standard carbureted gasoline engine to a dual-fueled gasoline/hydrogen or simply a hydrogen engine. The disadvantage is that carburetor engines are more susceptible to irregular combustion due to preignition and backfire problems. Power output from a hydrogen engine with a carburetor is about 15 percent lower than from a gasoline engine.

During the 1970s and 1980s, a number of techniques were researched to separate the fuel and air streams. At the University of Miami, researchers used experiments known as "hydrogen induction" to supply hydrogen. Hydrogen was supplied separately from the air through a tube close to the intake valve. They studied 19 various hydrogen engine configurations, with one advantage being their simplicity. As hydrogen engines continued to progress, direct injection into the combustion cylinder stroke became more common. Power output increased by more than 20 percent compared with gasoline engines.

Using dilution techniques, such as exhaust gas recirculation or water injection, alleviated preignition problems. Recirculation of 25 to 30 percent of exhaust gases eliminated backfiring. Nitrogen oxide emissions were also reduced by reducing the combustion temperature.

The overall thermal efficiency of a hydrogen engine is typically greater than that of a gasoline engine.

Another important parameter in engine performance is the power output. This is determined by converting the chemical energy of a fuel into mechanical energy. The power output of a hydrogen engine can be improved by advance fuel injection techniques or with liquid hydrogen. If liquid hydrogen rather than gaseous hydrogen is premixed, the amount of hydrogen that can be introduced into the combustion cylinder increases by approximately one-third, and the power output increases by approximately 31 percent. By directly injecting gaseous hydrogen into the cylinder under high pressure, the maximum amounts of both air and hydrogen introduced into the combustion chamber results in a 20 percent power increase compared with gasoline, and a 42 percent increase of power compared with a premixed gaseous hydrogen mixture.

Advantages also occur when hydrogen is mixed in controlled amounts with other fuels. With its fast flame speed, hydrogen can reduce the burning time of the primary fuel mixture, providing greater efficiency. Studies have found that hydrogen could be substituted for up to 38 percent of the total fuel mixture of diesel fuel with no loss of system efficiency and nominal loss of power output. Water injection increased the optimal substitution up to 66 percent. Smoke, exhaust, and nitrogen oxides were all reduced when water injection was used.

The greatest advantage of hydrogen vehicles compared with gasoline vehicles is their lack of pollutants. For a hydrogen engine, the principal exhausts produced are water and some nitrogen oxides. Small amounts of hydrogen peroxide have been released in inefficient hydrogen engines.

Nitrogen oxide emissions adversely affect air quality through the formation of ozone and acid rain. They are formed by all internal combustion engines today, no matter what the fuel. When the combustion temperature is high enough, nitrogen and oxygen from the air portion of the mixture combine to form nitrogen oxide. Nitrogen oxide production results from three variables: (1) the reaction temperatures, (2) the reaction duration, and (3) availability of oxygen. An increase in any of the three leads to an increase of nitrogen oxide. In hydrogen engines, nitrogen oxide levels can be reduced by running on a leaner mixture to reduce combustion temperatures. Also, nitrogen oxides can be reduced by cooling the combustion environment with water injection, with exhaust gas recirculation, or with cryogenic liquid hydrogen, which acts as a heat sink.

In summary, hydrogen can be used very efficiently in internal combustion engines. Early problems of preignition and backfire have been solved. Fuel injection directly into the cylinder and dilution techniques of recirculation of exhaust gas or water injection have been the answer. The major advantage of hydrogen as a fuel is its lack of pollution. The principal exhausts from a hydrogen engine are water and some nitrogen oxides. Hydrogen does provide a solution to the environmental dilemma.

USES FOR HYDROGEN

Today, the largest industrial application of hydrogen (about 95 percent) is as a chemical feedstock for production of certain chemicals. The three main industries include petroleum refining and resultant petrochemicals, ammonia manufacturing, and methanol production.

The petroleum industry is the largest consumer of hydrogen today, with hydrogen used in the refining process. Hydrogen removes sulfur from refinery products; it converts heavier oil fractions to lighter, more valuable oil fractions of fuels and lubricants, the process called hydro-cracking; and it is used in the chemical conversion of petroleum products. In hydro-cracking, hydrogen saturates the cracked products produced under the severe operating conditions of the hydro-cracking operation. Hydrogen is reacted with the petroleum fraction and then reacted with organic sulfur compounds to produce hydrogen sulfide, thus removing sulfur from refinery products.

The fertilizer industry consumes great amounts of hydrogen to produce ammonia, used in fertilizer-based products. In the Haber-Bosch process, a mixture of nitrogen and hydrogen in a 3:1 ratio passes over an iron oxide-promoted catalyst to produce ammonia.

Methanol (CH_3OH) production is the third largest consumer of hydrogen. Methanol is a major raw material of the organic chemical industry and is used as a feedstock for the production of formaldehyde resins (and subsequent wood laminate), dimethyl terephthalate (an essential product for polyester fibers), and solvents. Hydrogen is also a feedstock for the large synthetic plastics market.

Hydrogen is used for food processing in which unsaturated organic oils of soybeans, corn, peanuts, and coconuts are hydrogenated. Hydrogen stabilizes these oils and improves odor, color, and consistency. Inedible oils and greases are hydrogenated to produce soaps and animal feeds.

Hydrogen peroxide, hydrogen chloride gas, and organic chemicals all use hydrogen. The pharmaceutical industry uses hydrogen extensively to manufacture certain drugs. Hydrogen is also used in oxyhydrogen glass cutting and metal welding.

As we have seen and will discuss further, hydrogen is used as a fuel for internal combustion engines and turbines because of its high efficiency. It powers fuel cells, and fuel cells provide energy for electrical power generators and automotive power. The first stage of rocket propulsion uses hydrogen. Hydrogen can be used to power electric vehicles, air-conditioning systems, computers, refrigeration systems, and heat pumps. The current consumption of hydrogen is about 75 billion cubic meters per year, which is equal to about 10 percent of the total natural gas consumed in the United States. Hydrogen is an endless resource that can provide the world's energy needs indefinitely.

HISTORY OF HYDROGEN TRANSPORTATION

Early History

Evidence suggests that hydrogen fueled the first internal combustion engine. In 1820, Reverend W. Cecil developed a hydrogen-fueled engine based on the vacuum principle. Cecil burned hydrogen as a gaseous mixture which then expanded to atmospheric pressure and then cooled, resulting in a partial vacuum. The actual work was produced by the atmosphere, which forced the piston back against the vacuum. The engine ran satisfactorily, but vacuum engines were never practical. The technology was supplanted by engines that directly produced power on the expansion stroke, using increased pressure caused by combustion, the familiar internal combustion engine we use today.

In 1854, an English patent application was issued to Bursanti and Matteucci who developed a free-piston-type engine that burned hydrogen. In 1856, Benini built a prototype.

Otto engines, developed in the 1860s and 1870s, used substantial proportions of hydrogen in gaseous fuels, employed before the advent of gasoline. When Otto first experimented with gasoline, his results were so disastrous that he considered the fuel too unsafe and returned to the gaseous fuels. When the carburetor was developed, the use of gasoline became safe and practical.

World War I saw great advances with the use of hydrogen. Rudolf Erren, a German engineer, accomplished much of this both during and after the war. Erren converted a wide variety of engines to run on hydrogen. He promoted hydrogen engines for a variety of applications, including trucks, buses, and submarines. His engines could operate on either hydrogen or gasoline and could switch from one fuel to another or be mixed. By 1938, scores of "Errenized" trucks traveled between Berlin and the city of Ruhr, switching between the two fuels. Problems with these hydrogen engines included preignition and flashback, which Erren suggested solving by direct injection of hydrogen. The possibility of a completely condensible exhaust from a hydrogen-oxygen engine motivated his research on a hydrogen submarine. Most of Erren's research notes on hydrogen were destroyed in the Berlin bombings of World War II.

Airships

Beginning in 1900, great interest developed in the use of hydrogen as a primary fuel or fuel extender for the airships constructed in Germany and England. These airships were to be used for long-distance transportation of passengers and cargo. Hydrogen would provide the necessary buoyant force, but needed to be partially vented during the voyage. Storage would be no problem. Germany provided most of the research and development expertise in this area.

In 1900, Count Ferdinand Von Zeppelin designed and developed the first rigid airship called the "LZ-1." It successfully flew using hydrogen gas for aerostatic buoyancy. The first commercial air operation was started by a German transportation company using five Zeppelin airships. In the next five years, before operations were terminated by World War I, these hydrogen-filled airships made 1,600 flights carrying 37,250 passengers without injury. During World War I, many hydrogen-filled airships and barrage balloons were used. The Treaty of Versailles prohibited further development in Germany and prohibited the manufacture of airships by the Zeppelin Company. This prohibition was lifted when the U.S. Navy commissioned construction of the airship Los Angeles (LZ-126) as part of reparations. This airship was inflated with helium for buoyancy.

The next airship built by the Zeppelin Company was the LZ-127, the original Graf Zeppelin, completed in 1928. This 3,708,600 cubic-foot-rigid airship was decommissioned in 1937 after making 590 flights, including 144 ocean crossings, mostly between Germany and South America. It had flown 1,053,391 miles carrying 13,150 passengers and 235,300 pounds of cargo.

In 1936, the next Zeppelin, called the Hindenburg (LZ-129), was commissioned. It was 803 feet long and contained 7,063,000 cubic feet. The hydrogen-filled gas cells were contained within the cloth envelope that covered the aluminum framework.

Propellers helped the airship attain a maximum speed and turned a 4,110-horsepower Mercedes-Benz diesel engine. It could carry fifty passengers and had a range of 8,750 miles.

In 1936, the first commercial air service across the Atlantic took place between the United States and Germany. Eastbound crossings averaged 65 hours and westbound crossings took 52 hours aided by a tailwind. By May 6, 1937, the airship had crossed the Atlantic ten times carrying 1,002 passengers. On that fateful day in May at Lakehurst, New Jersey, the Hindenburg burst into flames and fell to the ground. Of the ninety-seven people onboard, including passengers and crew, thirty-five died. These were the first passenger fatalities in the history of commercial airship operation.

A discharge of atmospheric electricity in the vicinity of a hydrogen leak was recorded as the official cause of the fire. Two hours prior, an electrical storm had passed through the area. Some speculated that the fire was an act of sabotage initiated by antifascist sympathizers. Evidence of a photographic flash bulb was found in the wreckage, and one member of the crew killed in the accident had been involved in suspicious activity.

The crash of the Hindenburg signaled the end of hydrogen as the inflation medium of airships. Because of that singular occurrence, hydrogen received a bad reputation and was labeled an unsafe fuel, called the "Hindenburg syndrome." The ironic aftermath is that not one person was killed by the hydrogen fire. Passengers died because they jumped from the airship or died from the diesel fuel fire. The hydrogen storage cells were located above the passengers, and when the hydrogen burned, the flames shot up and away from the passengers. A recent investigation into the Hindenburg disaster concluded that the airship burned so vigorously, not because of the hydrogen, but because of the extremely flammable coating, a combination of iron oxide and aluminum. These chemicals are commonly used in rocket fuels.

Aircraft

In 1956, researchers undertook an experimental program to demonstrate the feasibility of burning hydrogen in an aircraft turbojet at high altitudes. A USAF B-57, a twin-engine medium bomber, was modified to carry liquid hydrogen tanks under the left wingtip. A similar size helium tank was carried under the right wingtip. Helium caused the liquid hydrogen to flow into a heat exchanger where it was vaporized. Scientists reported no safety problems during this successful testing.

Pratt and Whitney Aircraft Company received a contract to investigate the feasibility of using liquid hydrogen as a fuel in aircraft engines. Within five months, the company converted a J-57 engine to hydrogen. The research concluded that conventional jet engines could be readily adapted to use liquid hydrogen fuel. These efforts led Pratt and Whitney to develop the RL-10 hydrogen/oxygen liquid rocket engine for the U.S. space program.

U.S. Space Program

The first successful launch of a space vehicle was propelled by a liquid hydrogen/oxygen rocket engine on November 27, 1963, at Cape Canaveral. Kerosene (RP-1) and liquid oxygen fueled the Atlas stage rocket engine. This boosted the Centaur upper stage, which employed two of the RL-10 hydrogen/oxygen rocket engines developed by Pratt and Whitney.

In the rocket engine, liquid hydrogen circulated through the cooling jacket of the combustion chamber and nozzle assembly. It picked up enough heat so that when it expanded through the turbine, it provided enough power to pump the propellants to the pressure required to force them through the injection system into the combustion chamber. A small jet of hydrogen gas was used to remove the heat of friction from bearings and gears.

In all the Apollo launches, failure of a hydrogen-fueled rocket engine has never occurred. NASA gained a great amount of experience in handling liquid hydrogen. This included experience with manufacturing, liquefaction of hydrogen, transportation, storage, pipelines, instrumentation, design practices, operational use, and safety procedures.

Hydrogen Vehicles

U.S. History: Roger Billings, considered to be the U.S. pioneer of hydrogen vehicles, made the first converted hydrogen-fueled internal combustion engine in the United States. In the mid-1960s, at the age of sixteen years old, he converted a Model A truck. Billings also experimented with converting lawn mowers to run on hydrogen. In the 1970s, Billings began a concerted effort to build hydrogen-powered vehicles when he teamed up with Frank Lynch, who had been a student member of a UCLA team that designed a hydrogen-powered car. Billings Energy Corporation resulted, and they converted a number of vehicles to hydrogen throughout the 1970s and 1980s. The pair also developed an onboard metal hydride tank, an iron-titanium storage container that utilizes hot water for releasing hydrogen gas. The team also converted a Monte Carlo car and equipped it with both cryogenic hydrogen and metal hydride storage.

The Billings Energy Corporation converted two buses to run on hydrogen. One bus, delivered to Riverside, California, operated in regular service by the local transit authority. Billings outfitted a U.S. Post Office Jeep with a carburetor modified to accommodate hydrogen. He utilized a water injection system to suppress backfire, which helped the Jeep obtain 21 percent greater fuel efficiency compared with a gasoline-fueled Jeep. Billings also constructed a prototype hydrogen fuel cell vehicle, a Ford Fiesta that had been retrofitted as an electric vehicle by the U.S. Post Office. Hydrogen was stored onboard in a metal hydride system.

Frank Lynch founded a company called Hydrogen Consultants, Inc. (HCI). For several decades, Lynch modified vehicles for hydrogen operation. In the early 1980s, the company investigated the feasibility of using hydrogen for underground mining machinery. He converted a Caterpillar diesel engine to a hydrogen-fueled spark-ignition engine, with fuel delivered by a parallel injection technique. He performed a similar conversion on an eighteen-passenger Mitsubishi Rosa bus. Hydrogen Consultants played a major role in the development of the Riverside, California hydrogen program.

Energy Partners of Palm Beach, Florida, built a prototype fuel cell electric vehicle called the "Green Car." It was powered by three Nafion Proton Exchange Membrane (PEM) fuel cell stacks. The car also had a 3,000-psi storage cylinder and batteries for cold starts and for extra acceleration. The car reached speeds of 60 mph, and it had a range of sixty miles for city driving. In ten seconds, it was able to accelerate from 0 to 30 mph.

H-Power was the prime contractor for the Department of Energy phosphoric acid, fuel cell bus project. The company delivered three fuel cell buses, unveiled in April 1994. They developed another bus based on the PEM fuel cell technology, designed as a retrofit for existing diesel buses. It used a reformer to convert diesel fuel to hydrogen, enabling it to refuel with diesel fuel. H-Power also developed the sponge-iron hydrogen storage technology.

Ben Jordan, another U.S. hydrogen pioneer, has operated at least a dozen engines on some form of hydrogen. He helped organize races for hydrogen-powered vehicles at the Bonneville Salt Flats in Utah.

Los Alamos National Laboratories (LANL) has researched several hydrogen vehicle projects and has been involved in an extensive fuel cell research program. The laboratory converted a 1979 Buick Century to operate on liquid hydrogen and tested it for 2.5 years. The car was driven 2,252 miles and operated for 138 hours. When compared with a gasoline car, it had less power and some backfire problems. Two liquid hydrogen storage tanks were tested on the car. A 29-gallon tank had a range of 170 miles, and a 40-gallon tank had a range of 224 miles. The car refueled sixty-five times with an average refueling time of nine minutes.

The Los Alamos National Laboratories also researched a fuel cell electric golf cart and concluded that a fuel cell vehicle could operate at greater efficiency than comparable performance with a gasoline vehicle.

Researchers at Los Alamos are developing a hydrogen refueling station to accommodate liquid hydrogen. They are researching refueling with medium-pressure gas and refueling with low-pressure gas that is required in metal hydride storage systems.

In 1987, Dr. Cliff Rickets, at Middle Tennessee State University, operated the university's first hydrogen engine. In 1989, the university team had the fastest time for a hydrogen-fueled vehicle, 88 mph at the Bonneville Speed Flats. By 1994, the team's vehicle reached 108 mph. Two of the valves used in the vehicles supplied hydrogen and air separately to the combustion chamber so that the engines could operate more efficiently without premixing the hydrogen and oxygen.

Robert Zweig, M.D., founded the Clean Fuels Institute (CFI) that acquired nineteen hydrogen passenger buses for the City of Riverside, California. The Billings Energy Corporation built these buses, which featured a ten-vessel iron-titanium hydride storage system that was heated by circulating engine coolant through a water jacket on the tanks. The buses, also equipped with propane fuel, could switch between propane and hydrogen with the flip of a switch. Backfiring and inadequate power were some of the earlier problems encountered.

The University of California at Riverside College of Engineering experimented with two hydrogen vehicles. A Ford Ranger stored hydrogen gas onboard in high-pressure gas cylinders made of composites. Hydrogen was produced at the on-site solar electrolysis facility. Students also developed a small electric vehicle with an onboard hydrogen fuel cell and a hydride storage system. The hybrid electric vehicle used the electricity from a fuel cell to keep the batteries charged, using a 1.5-kilowatt phosphoric acid fuel cell. Hydrogen gas was stored at 4,500 psi in lightweight composite pressure containers.

Throughout the 1970s, UCLA was involved in a number of projects to develop and test hydrogen vehicles. Professor William VanVorst initiated the program, which converted American Motors Gremlins to cryogenic hydrogen operation.

The University of Miami has one of the most important hydrogen engine development programs in the nation. Miami researchers were among the first to investigate advance delivery systems as a means to eliminate premature ignition in a hydrogen engine. This led to the hydrogen induction technique.

From 1977 to 1982, the University of Miami studied various engine configurations and provided a design database for light-duty hydrogen-fueled engines. They tested nineteen various engine configurations. Researchers also tested hythane, a mixture of natural gas and hydrogen. Another project involved redesigning the combustion chamber as opposed to the fuel delivery system to accommodate hydrogen fuels.

T. Nejat Veziroglu, Ph.D., from the University of Miami, is one of the world's leading proponents of hydrogen energy. In 1974, he organized the first major conference on hydrogen energy, and now annual international meetings involve many countries. Veziroglu was nominated for the Nobel Prize in Economics in the year 2000 for his work with the hydrogen economy.

Roy McAlister, another hydrogen pioneer, founded the American Hydrogen Association. McAlister, a mechanical engineer, has developed several patents regarding hydrogen technology. He provides workshops and accompanying kits on converting the internal combustion engine to run on hydrogen, with the option of choosing the fuel type while operating the vehicle. He believes we should make the transition to the hydrogen economy as quickly as possible.

Australia: Scientists at the University of Melbourne have been researching hydrogen fuel for transportation since the 1970s. They developed a hydrogen-fueled vehicle powered by a two-liter, four-cylinder Ford engine equipped with a delayed port admission fuel system. When compared with a gasoline engine, the vehicle had low emissions and a 13 percent improvement in fuel efficiency.

Belgium: The Euro Quebec Hydro Hydrogen Pilot Project (EQHHPP) was designed to show the feasibility of hydrogen transportation to promote the use of hydrogen in Europe. The first four buses constructed under EQHHPP are being demonstrated and tested in Belgium. The eighty-eight passenger Van Hool bus, called the "Green Bus," was converted to run on liquid hydrogen by VCST-Hydrogen Systems of Belgium. It is powered by a six-cylinder diesel engine and equipped with a 125-liter liquid hydrogen tank. No carbon monoxide, carbon dioxide, or hydrocarbon emissions have been detected, and nitrogen oxide emissions have been reduced by 2,800 percent. Disappointingly, the bus has a range of only 37.5 miles and a top speed of 39.3 mph.

Brazil: The UNICAMP laboratory in Brazil has been involved in hydrogen research for more than twenty years, producing two hydrogen vehicles, a Toyota and a Kombi-Volkswagen van. The Toyota runs on a mixture of hydrogen and diesel fuels, with 20 percent of the energy value coming from hydrogen. Operating on pure hydrogen, the Kombi-Volkswagen van uses iron-titanium and manganese-nickel metal hydride tanks. Researchers experienced no difficulty with the Toyota, but the van experienced large power losses and backfire problems.

Canada: Ballard Power Systems is currently demonstrating a transit bus system powered by PEM fuel cells. A 32-foot-long bus uses twenty-four Ballard 5-kilowatt PEM fuel cell stacks arranged in three parallel rows of eight to provide a gross power of 120 kilowatts. The fuel cells can deliver all the necessary power to run the bus. It has been driven more than 1,200 miles between Vancouver, Sacramento, and Los Angeles. Ballard developed a second hydrogen transit bus that was developed with a range of 240 miles. A third fuel cell bus introduced in1998 has a range of 330 miles.

The Eureka Project: A joint venture among Belgium, the Netherlands, France, and Italy was initiated in 1992 to research fuel cells. The goal was to demonstrate an alkaline/hydrogen fuel cell system as part of a hybrid electric system in a city bus. Nickel-cadmium batteries provided a secondary source of power. Liquid hydrogen tanks were located on the rear of the bus.

Germany: Since 1973, Daimler-Benz (now DaimlerChrysler) has researched the possibilities of using hydrogen as a fuel for surface transportation. Beginning in 1984, the company created a fleet of vehicles in Berlin, composed of two types of vehicles. They developed five models of 280TE passenger cars that used a mixture of hydrogen and gasoline. Also, five models of 310 delivery vans were converted to be used with pure hydrogen fuel. Both vehicle types stored hydrogen in a metal hydride system. The vehicles operated reliably and safely during the four-year test period. Because of the heavy weight of the hydride tank, the tires, brakes, and springs showed signs of increased wear. In 1989, a more advanced model 230E was developed with a modified intake system to mix hydrogen and air before injection into the engine cylinder. Hydrogen was stored in a metal hydride system weighing 700 pounds that allowed a range of 80 miles.

DaimlerChrysler is involved in a Swiss/German cooperative program regarding investigating a hydrogen application that uses seasonal and weekly surpluses of electricity. The program built a prototype hydrogen bus fueled by pressurized hydrogen tanks mounted on the roof.

DaimlerChrysler introduced a Mercedes van powered with a fuel cell developed jointly with Ballard Power Systems of Canada. The company announced that in 2003, it will introduce a hydrogen version of the Dodge Durango.

Another major researcher of hydrogen vehicles, BMW, developed a liquid hydrogen prototype. The BMW 750i operates on liquid hydrogen with a turbocharge, fueled by timed individual post injection. A later chapter will discuss BMW's current research.

Japan: The Musashi Institute of Technology began to study hydrogen engines in 1970. A series of hydrogen vehicles now bear the name of Musashi. A one-ton truck was their first vehicle (Musashi 1) equipped with high-pressure hydrogen tanks. Musashi 2 has a 230-liter liquid hydrogen storage dewar in the rear of the car. Liquid hydrogen also fueled the Musashi 9, a refrigerator van that used the liquid hydrogen to cool the refrigerator to 0 degrees centigrade (32 degrees Fahrenheit) in addition to fueling the engine.

Mazda has developed several prototypes based on their rotary engines. The engine style is ideal for hydrogen because the operational chamber moves so that intake, compression, combustion, and exhaust occur at different places. Backfire and preignition can be reduced because the temperature during intake is low, and no direct contact occurs between the air/fuel mixture and exhaust gases.

In 1991, Mazda introduced its first hydrogen concept car, the HR-X, at the Tokyo Motor Show. Features include a metal hydride storage tank and active torque control system to provide extra power. This system uses a battery-powered electric motor to add additional torque to the engine during acceleration periods.

When braking, the system converts excess momentum into electricity, which is stored for later use in a nickel-metal hydride battery. An updated HR-X2 incorporates a larger displacement engine with a peripheral port for enhanced power. An improved metal hydride tank is composed of eighteen individual rectangular cells.

Mazda has also released a hydrogen-fueled Miata and a 626 station wagon. The Miata has a cell-type metal hydride similar to the HR-X2. The company has also developed a fuel cell-powered golf cart.

Former Soviet Union: Two institutions were the principal centers of hydrogen research in the former Soviet Union: (1) the Institute for Problems in Machinery-Ukrania Institute of Automobiles, and (2) the Research Institute of Automobiles-Ministry of Transportation. Both institutes have experimented with dual fuels, combining gasoline and hydrogen, in cars, vans, and trucks. Their test results showed gasoline consumption decreased 38 percent and fuel economy increased 21 percent when vehicles operated in a dual mode as compared with gasoline-only operation. Significant reduction in the emissions of carbon monoxide, hydrocarbons, and nitrogen oxides occurred under dual fuel operation.

This brief history of hydrogen vehicles has highlighted a number of problems regarding hydrogen powered vehicles. Earlier researchers encountered problems with preignition, flashback, onboard storage, power density, and driving range. Current research has addressed these problems, and later chapters will update the reader on the present-day status of hydrogen technology and the problems that still need to be solved.

Chapter Three

THE HYDROGEN INFRASTRUCTURE

Production, Storage, and Distribution

The technology is available today to convert to a hydrogen economy, but a transition period is required to completely wean ourselves from fossil fuels. Fossil fuels are carriers of hydrogen, and during the transition period, they will be a major source of hydrogen, whether from coal, oil, or natural gas, the latter being more environmentally friendly. Water contains two atoms of hydrogen, and it takes a significant amount of energy to extract hydrogen from water. To produce one kilogram of hydrogen from water, it takes an equivalent of burning five kilograms of coal.Oil companies will need to be the suppliers of hydrogen energy because they have an infrastructure already in place. In the future, water will provide the major source of hydrogen through electrolysis. Currently, electrolysis of water is quite expensive and requires fossil fuels to provide the electricity. During the transition period, a new infrastructure will develop that relies on renewable energy from the sun. Eventually, the production of electricity will be from solar thermal energy, photovoltaics, wind energy, hydroelectric, and ocean energy. The solar energy supply is unlimited and will eventually produce hydrogen by electrolysis.

Hydrogen, whether in the form of gas or liquid, must be stored. Compressed hydrogen gas can be stored in containers that have more versatility than liquid hydrogen containers. Liquid hydrogen has to be kept at a very low temperature to maintain its liquidity. The advantage of liquid hydrogen is that more energy is compacted into a smaller volume compared with the gaseous state, providing a greater range for transportation vehicles. The third method for storing hydrogen is in metal hydrides, in which hydrogen adheres to the surfaces and crevices in finely shaved metal. Once heated, the metal releases the hydrogen that is converted into energy.

Distribution of hydrogen to the consumer can occur through the existing fossil fuel infrastructure. Hydrogen produced at natural gas facilities can use the existing pipelines that supply natural gas to customers around the country. Trucks can transport liquid hydrogen and compressed hydrogen to refueling stations, similar to today's gasoline distribution system.

The one great advantage of hydrogen is that it can be produced on-site, an occurrence that will come about more frequently during the transition period. Communities, industries, and private individuals will have the option of producing their own hydrogen with renewable means. The existing infrastructure is in place to begin the transition to a hydrogen economy.

HYDROGEN PRODUCTION

Steam Reformation

Steam reformation is the most widely used technique for hydrogen production today, being the most efficient and economical. It basically involves converting hydrocarbons (such as natural gas) and steam to hydrogen, carbon dioxide, and carbon monoxide. The four-step process involves the following:

1. purification of the feedstock to remove sulfur; and carbon oxides;
2. steam reforming of the hydrocarbon to form hydrogen and carbon oxides;
3. converting carbon monoxide to carbon dioxide; and
4. final purification to remove carbon dioxide, carbon monoxide, and remaining hydrocarbons.

Only light hydrocarbons, which can completely vaporize without leaving a carbon residue, are used. Carbon residues can hinder the catalysts in the steam reforming process. Natural gas is the preferred feedstock for hydrogen production. Other feedstocks include naphtha, and Number 2 fuel oil. Natural gas is easy to handle, has the highest hydrogen to carbon ratio (4:1), and generally has low levels of sulfur contamination. Because natural gas reacts more easily with water than with other hydrocarbons, lower temperatures can be used for the conversion.

The steam reformation process uses a large amount of energy and high pressure. It requires a temperature from 760 to 925 degrees centigrade, at a pressure twenty times that of atmospheric pressure. Catalysts are used to improve the efficiency of the reaction. They are usually compounds of nickel oxides mixed with oxides of aluminum, calcium, manganese, or other minerals.

Steam reformation is reversible, meaning hydrogen can convert back to methane. To ensure a minimum concentration in the reformed gas, the process usually employs a steam to carbon ratio of 3:5 at processing temperatures of 815 degrees centigrade and pressures up to 35 times atmospheric pressure. This allows for 80 percent of the hydrocarbons to be converted to carbon oxides. Energy intensive, the steam reformation process accounts for a significant amount of the total energy usage. In other words, energy is needed to manufacture hydrogen.

After the reforming process, the gas mixture is fed through a heat recovery step and reacted with excess steam. This converts carbon monoxide to carbon dioxide, reducing the carbon monoxide to less than 1 percent. The remaining carbon monoxide and carbon dioxide can be removed using a methanation reaction, leaving a hydrogen purity of 97 to 98 percent, which is adequate for most uses as a fuel. To reach 99.97 percent purity, a pressure swing adsorption technique is used, passing the raw gas through a series of activated carbon beds.

Partial Oxidation

Partial oxidation is similar to steam reformation in the production of hydrogen. The process involves steam, oxygen, and hydrocarbons that are converted to hydrogen and oxides of carbon. The process of partial oxidation requires additional facilities to separate oxygen from nitrogen in the air and requires separation equipment for the conversion and removal of carbon dioxide. Compared with steam reformation, an additional step is required, where oxygen is used to oxidize hydrocarbons to provide sufficient energy to drive the process and release additional amounts of hydrogen. Feedstocks range from methane to coal.

Coal Gasification

In their ninety gasification plants, Texaco has been the leader in developing coal gasification technology. From a single ton of Kentucky Number 9 coal, almost 50,000 cubic feet of hydrogen can be produced. This process is the cleanest commercial method of coal utilization today. Sulfur is removed and transformed in the process; it is then sold as an agricultural chemical. A resultant dry synthetic gas is saturated with water to suppress nitrogen oxide emissions.

Texaco's Cool Water Coal Gasification Program plant is located on 195 acres midway between Los Angeles and Las Vegas, with an adjacent 103 acres used for railroad loops, coal handling, and storage. In this plant, coal is converted to a synthetic gas that has approximately one-third the heating value of natural gas. The synthetic gas is routed into a catalytic water solution for a reaction. An oxygen atom is removed from a water (steam) molecule and added to a carbon monoxide molecule to make carbon dioxide. Therefore, each carbon monoxide molecule becomes a carbon dioxide molecule, and each water molecule becomes a hydrogen molecule. This leaves a mixture of hydrogen and carbon dioxide that is routed to an acid-gas removal section where the carbon dioxide is removed. The resulting gas steam is primarily hydrogen. The clean synthetic gas is fired in a combustion turbine, and the resultant heat produces steam. This steam drives a turbine. The net electricity produced from this plant is equivalent to about 4,300 barrels of oil per day, producing 117 megawatts and serving the energy needs of 100,000 households. This product is equivalent to the 70 million cubic feet per day of hydrogen produced by a 1000-ton-a-day coal plant.

Texaco's cool water program has received worldwide recognition. Emissions of sulfur oxides, nitrogen oxides, and particulates average about 10 percent of the maximum levels permitted by the EPA for new coal fuel electric generating plants. Texaco believes coal gasification is the preferred option for hydrogen production plants on a long term basis. The Japanese Ube Ammonia Industry Company uses Texaco's gasification system to convert 1,650 tons of coal per day into hydrogen for the manufacture of 1,000 tons of ammonia per day.

Steam-Iron Process

The steam-iron process was commonly used to produce industrial hydrogen before the inexpensive natural gas steam reformation process was developed. This technology involves steam being reacted with hot iron to produce gas at a temperature ranging from 815 to 870 degrees centigrade. The first step involves coal gasification that uses steam and regular air, producing a mixture mostly of carbon monoxide and hydrogen. The mixture is introduced into a second chamber with iron, producing iron oxide (FeO). The third step involves the transfer of the hot iron to a steam-iron reactor. The iron oxidizes in the presence of steam to produce an iron oxide (Fe_30_4) and a hydrogen-rich gas. The gas mixture is purified by condensing out the steam and running the remaining gas through a methanation reaction, resulting in 97 percent pure hydrogen gas. This process requires a great amount of power.

Water Electrolysis

To produce hydrogen electrolytically requires only pure water and electrical energy. In an ideal world, nonpolluting renewable resources such as solar thermal energy, photovoltaics, wind energy, or hydroelectric power would generate the needed electricity. Current renewable techniques are not economically feasible for large-scale hydrogen production. Electric energy from fossil fuels for electrolysis is expensive, with costs three to five times greater than hydrogen production from fossil fuels.

In electrolysis, electricity decomposes water into its constituents, gaseous hydrogen and oxygen. Two electrodes are used in electrolysis, one positive and one negative. They are immersed in a current-conducting electrolyte that may consist of a mixture of water with a salt, acid, or base. The most common electrolytes used are alkaline electrolytes such as potassium hydroxide (KOH). These electrolytes conduct electricity very well and are quite stable, with minor corrosion problems.

The positively charged potassium ion (K+) is reduced and reacts with the water to form a hydrogen atom (H) and a hydroxyl atom (OH). The highly reactive hydrogen atoms bind to the metal of the negative cathode. These atoms can combine with one another to form hydrogen molecules (H_2), which leave the electrode as a gas. At the positive anode, the hydroxyl atoms attach to one another, producing water and oxygen. Two oxygen atoms combine to form an oxygen molecule.

As technology advances, the efficiency of water electrolysis continues to improve. Three promising advanced technologies are under investigation:

1. Advanced water electrolysis involves operating the cells at temperatures greater than 100 degrees centigrade. This improves voltage efficiency, decreasing the amount of energy required to produce hydrogen.
2. Solid polymer electrolysis technology involves a solid rather than a liquid electrolyte. Its compactness and simplicity make it a viable technology for commercial electrolysis. Nafion is the solid polymer used. When saturated with water, the polymer becomes acidic and conducts ions. Nafion is presently very expensive since it requires high amounts of platinum catalyst loading.
3. High temperature electrolysis uses a solid electrolyte that does not corrode. It decreases the need for electrical energy and minimizes over voltage. The materials are costly, and the process requires a high temperature heat source.

Thermal Decomposition

Thermal decomposition (pyrolysis) of natural gas can produce hydrogen. At temperatures above 900 degrees centigrade and pressures up to 150 psi (150 pounds per square inch or 10 atmospheres), methane almost completely decomposes to carbon and hydrogen. Energy required for methane decomposition is less than that required for steam reformation of methane. However, the thermal efficiency for steam reformation is 75 percent higher than for methane decomposition. On the positive side, no carbon dioxide emission occurs in the thermal decomposition of methane by pyrolysis.

A low-cost process for hydrogen production involves thermal decomposition of water vapor at temperatures up to 2,800 degrees centigrade using solar energy and oxygen in high-temperature ceramic membranes. Avoiding conversion of solar energy to electrical energy minimizes the energy required to produce hydrogen.

Biomass Gasification

Technology to produce hydrogen by biomass gasification with steam and oxygen has been progressing. Biomass, such as plant matter, uses solar energy (photosynthesis) as the primary energy source to produce hydrogen, thereby becoming a feedstock. Biomass is defined as all plant and animal material that can be converted into energy. Biomass comes from a variety of sources including residues, wastes, and crops grown for energy use. Plants used for energy production include crop residues, forest residues, standing vegetation, and energy crops. Growing biomass by rapid crop rotation for wood production is one type of energy farm.

Animal waste, mainly from livestock, can be used as an energy source in the form of manure. Municipal solid wastes are also used as a feedstock for hydrogen, with the organic fraction extracted by size reduction and separation processes. Industrial wastes generated by food processing plants such as canneries, breweries, and sugar mills can produce hydrogen. Even the organic fraction of sewage wastewater can be used after it is separated and dried.

Modern technologies can convert biomass into a variety of products including methane, methanol, oils, charcoal, or directly into hydrogen. Through steam reformation or gasification, hydrogen can be extracted from all these hydrocarbons.

In the United States, energy from biomass could supply an estimated equivalent of 7 million barrels of oil per day. Waste resources could supply 40 percent, and energy crops could supply 60 percent, grown on 80 million acres of excess agriculture land. The crops of sugarcane and sorghum on 66.4 million acres that were taken out of U.S. production between 1981 and 1988 could have supplied enough energy to generate 34 percent of the nonnuclear power generated in 1986. In other words, biomass could supply a great amount of hydrogen.

Hydrogen From Bacterial Sources

Hydrogen can be acquired from bacteria. Cyanobacteria have two important hydrogen metabolizing enzymes. Nitrogenase produces hydrogen molecules and hydrogenase consumes them. To produce hydrogen, Cyanobacteria must be kept in an anaerobic (oxygen-free) environment. Otherwise, they would consume their own hydrogen. They are ideal microorganisms for producing hydrogen because of their ability to perform photosynthetic carbon dioxide fixation, and because they can thrive in a simple inorganic environment.

Photosynthetic bacteria can produce hydrogen from organic acids, alcohols, sugars, and a few sulfur compounds at a high rate for longer periods than can Cyanobacteria.

Anaerobic digestion is used to produce methane gas and other useful residues from a variety of carbonaceous materials. The waste from one set of organisms is the food for another set. Digestion takes place in the presence of anaerobic bacteria. Cellulose and hemi-cellulose are broken down by enzymes to form soluble organic compounds. These compounds are converted to hydrogen, carbon dioxide, and acetate by acid-producing bacteria. Methane is produced by methanol-producing bacteria from the hydrogen, carbon dioxide, and acetates.

Stirred tanks are the mainstays of most digestion systems. Anaerobic digestion is the preferred process for gasification of high-moisture grains. The resultant gas (mostly methane) has a medium fuel value. Hydrogen can be obtained from the gas through steam reformation, but the additional process is not cost effective at present.

Renewable Sources

Hydrogen is already produced in great quantities today, about 500 billion cubic meters annually. However, only 1 percent is produced from solar, wind, or water. The predominant source (about 50 percent) is natural gas.

A breakthrough has occurred at the National Renewable Energy Laboratory (NREL), headquartered at Golden, Colorado. NREL recently announced the invention of a semiconductor device that converts light into hydrogen that is suitable for a fuel. In one step, light goes in and hydrogen comes out. NREL claims the device has an efficiency twice any previous device that converts light energy to hydrogen. Scientists are curious to examine the technology underlying this remarkable claim.

HYDROGEN STORAGE

Until this time in the hydrogen story, storage has presented a major hindrance in the progression toward a hydrogen economy. Hydrogen has a low energy density, meaning that it requires a considerably greater volume for storage than does gasoline. This translates into larger storage tanks and shorter driving ranges for vehicles. Consumers are used to traveling great distances before refueling their vehicles. Rapidly developing technology promises to extend these driving ranges to make them comparable with those of gasoline vehicles. Three major storage systems exist for hydrogen: compressed hydrogen gas, liquid hydrogen, and metal hydrides.

Compressed Gas

Because of hydrogen's low energy density, a compressed hydrogen gas tank at one time needed to be twenty times larger than a comparable gasoline tank. This figure has been reduced as the compression tank technology has progressed.

The high-pressure vessels used to store hydrogen are typically aluminum cylinders wrapped with carbon fibers in fiberglass. Other designs being researched include aluminum wrapped with Kevlar or graphite, plastic liners wrapped with graphite, and high-strength aluminum cylinders. Both volume and mass are issues for storing hydrogen as a compressed gas. A pressurized hydrogen system at approximately 3,000 psi weighs nearly three times more than a comparable liquid hydrogen system. The volume can be reduced by increasing the pressure of the compressed gas, potentially up to 8,000 psi. For these high-pressure vessels, cost and safety become an issue. Compressing the hydrogen gas to these pressures requires more energy.

At one time, a hydrogen vehicle with a 100-mile range, compared with the range of an electric vehicle, would require two hydrogen tanks weighing about 100 pounds. Such a vehicle would lose about one-half of its trunk space. A hydrogen vehicle with a 200-mile range is also possible, but the trunk space would essentially be eliminated by adding two more tanks. Higher pressure storage vessels have extended the driving ranges.

As we will see in the next chapter, a more efficient fuel cell will expand the range of a hydrogen vehicle. Storage needed for such a hydrogen fuel cell with a range of 250 miles is equivalent to that of an internal combustion engine with a range of 161 miles.

After hydrogen gas is produced, it is conventionally stored in banks of "K" bottles or "tube" tanks, at pressures equaling 6,000 psi. Tube tanks are fabricated from low-carbon steel, and can be used at pressures as high as 10,000 psi. Larger quantities of gaseous hydrogen can be stored in laminated vessels at pressure levels of 15,000 psi. These high pressure containers are double walled, and the inner layer is compatible with hydrogen at the higher stress levels. The outer wall is made of high-strength welded steel to contain the high pressure. During high levels of hydrogen production, hydrogen can be stored in the pipeline system at elevated pressures.

Recent advances in composite vessels have improved the efficiency of compressed storage systems significantly. They have been designed to hold much higher storage pressures, achieving volumetric efficiencies comparable to cryogenic systems.

Composite vessels consist of a thin metal liner and a high tensile fiber over-wrap. The main function of the liner is to prevent leakage. The over-wrap consists of carbon aramide fibers (ten times stronger than steel) that exhibit a very high tensile strength despite its low specific weight. There are two main types of cylinders.

The hood-wrapped cylinder has a metallic liner designed for a pressure of 100 bar (1,500 psi), which is half the operating pressure (3,000 psi) for the composite cylinder. The over-wrap is made of a fiber reinforced plastic that enables the cylinder to be operated at service pressure.

A fully-wrapped cylinder contains a metallic liner designed for an operating pressure of approximately 600 psi. The metallic liner may be replaced with a plastic liner that serves as a barrier to prevent gas diffusion.

In the Munich Airport hydrogen project, lightweight composite tanks are used as the pressurized vessels for storing liquid hydrogen. They have an aluminum liner and a epoxy/carbon fiber compound winding. The low floor buses accommodate a total of fifteen pressurized gas tanks that are distributed fore and after body. These lightweight tanks have a bursting pressure of 9,000 psi.

Undergound Storage

Two categories of underground storage facilities exist. Hydrogen gas can be pumped into depleted oil and gas fields or into aquifers. Called the "porous media storage system," hydrogen gas occupies the naturally occurring pore spaces between mineral grains. Hydrogen gas can also be contained in excavated or mined cavities in dense rock, called "cavern storage." In England, hydrogen has been stored in solution in mined salt caverns. It is stored at 750 psi at a depth of 1,200 feet. In France, hydrogen gas was stored in an aquifer near Beyenes, France. Underground storage of hydrogen is an inexpensive method, especially in depleted oil and gas fields that require a low capital investment.

Carbon dioxide can also be stored underground. In Norway, the energy company Norsk Hydro is constructing a power plant that will be fueled with hydrogen acquired from natural gas. The remaining carbon dioxide will be injected into an oil field on the continental shelf. The carbon dioxide will be taken out of circulation and will pressurize the oil field, allowing the remaining oil to be pumped easier. In Europe and the United States, pumping carbon dioxide into underground aquifers has proved an effective way of keeping it out of the atmosphere.

Liquid Storage

The boiling point of hydrogen presents the greatest difficulty in using a liquid hydrogen system. The boiling point is -423 degrees Fahrenheit (-253 degrees Centigrade) at atmospheric pressure. To maintain hydrogen in a liquid form, an extremely low temperature must be maintained. Modern technology in cryogenics has solved this problem. State-of-the art NASA technology uses liquid hydrogen for rocket propulsion. Through research at Los Alamos National Laboratories, a liquid hydrogen delivery system has been developed. As we will see later, BMW in Germany operates its prototype hydrogen cars on liquid hydrogen.

By weight, liquid hydrogen has the highest energy density available in a chemical fuel. For this reason, liquid hydrogen is the principal fuel used by NASA for rockets. NASA has built spherical vacuum-jacket containers that can hold 850,000 gallons of liquid hydrogen, thus ensuring a continuous supply.

The liquefaction process is energy intensive and involves several steps. First, gaseous hydrogen must be compressed. Next, the compressed gas must be precooled with liquid nitrogen at temperatures of -382 degrees Fahrenheit (-230 degrees centigrade). It then must be expanded through turbines and then converted from ortho- to para-hydrogen spin isomers. The latter step helps prevent large boil-off losses of hydrogen when cooled from its normal state. The energy requirement for the liquefaction process is the equivalent of 40 percent of the combustion energy of hydrogen itself.

A variety of intermediate liquids can be used to store hydrogen. Some carriers are called "open-loop carriers," which generate an end product that is not recyclable. Examples are ammonia (NH_3) and methanol (CH_3OH), which can be stored as liquids and used for hydrogen production as needed.

To produce hydrogen from an ammonia carrier, ammonia vapor is exposed to an iron oxide type catalyst at temperatures exceeding 700 degrees centigrade to yield hydrogen ($2NH_3 = N_2 + 3H_2$). Steam reformation techniques to produce hydrogen in large-scale plants can be used to release hydrogen stored in methanol ($CH_3OH + H_2 = CO_2 + 3H_2$).

A "closed-loop carrier system" consists of liquid hydrogen that can be stripped of some or all of its hydrogen. The resultant products can be saved for subsequent recharge with hydrogen. Hydrogen stored in these liquids can be transported to a central station where they are dehydrogenated. The CNR Institute in Messina, Italy, has worked with benzene and cyclohexane carriers.

Liquid hydrogen has been researched extensively in Japan, Germany, and the United States. To maintain the low temperature required, liquid hydrogen is commonly stored in double-walled superinsulated vessels. Hydrogen is withdrawn from these vessels in either a liquid or gas phase to be delivered to the engine. These insulated vessels help minimize the transfer of heat from the outside element, thereby reducing the boil-off hydrogen. Evaporation rates of these vessels are about 2 percent per day or less.

A liquid hydrogen system weighs slightly more than a comparable gasoline system, but is considerably more bulky. Hydrogen carried in cylinders requires 3.75 times more space than does gasoline.

The advantage of delivering liquid hydrogen to the engine cold (-80 degrees centigrade) is that the efficiency of the engine can be increased by producing a higher energy charge. It also reduces the temperature in the combustion chamber, thus reducing nitrogen oxide emissions.

Metal Hydrides

Metal hydrides offer another storage system for hydrogen, based on the concept that gaseous hydrogen is readily absorbed into metals, forming a weak chemical bond. To release hydrogen from the metal, the hydride is heated to a certain temperature. The exhaust heat from the engine, carried by water or exhaust gases, is commonly used for this purpose.

The advantage of metal hydride systems is that they are easier to deal with than other onboard storage systems. They do not require high-pressure tanks or cryogenic temperatures to store hydrogen. Hydrogen from a hydride storage system does not have enough pressure for use in a high-pressure injection application.

Metal hydride systems are one of the safest alternatives for storing hydrogen.The greatest disadvantage is that they have a low mass energy density, and thus they tend to be very heavy. A hydride storage system weighs about ten to twenty times more than a gasoline tank for the same energy weight. This increased weight limits the driving range from 93 to 186 miles. Hydride systems can be very large in volume, usually ranging from 25 to 75 gallons. In addition, contamination by oxygen or water results in reduced storage capacity, sometimes as much as 2 percent after two refueling cycles.

Two types of metal hydrides exist, low temperature and high temperature. A ferro-titanium system has great promise as a low-temperature hydride. It has good absorption capabilities at low cost. When manganese is added to the system, the lattice expands, leading to stabilization of the hydrogen phases and decreased storage pressure. The maximum capacity for hydrogen in a ferro-titanium-manganese system is about 1.9 percent for the total mass at 300 psi and a temperature between 32 degrees to 68 degrees Fahrenheit. Hydrogen can be released from the hydrides using only heat from the exhaust. Approximately 1,400 pounds of the hydride would be required to fuel an automobile with a cruising range of 300 miles. The volume required would equal 90 gallons, excluding the equipment.

High-temperature hydrides are usually lighter, compared with lower temperature hydrides. The weight of the overall system can be reduced by using a combined system, that is, a high-temperature hydride used with a low-temperature hydride. The optimum system appears to be a magnesium-nickel system, with nickel speeding the reaction. Magnesium-based hydrides offer the advantage of higher hydrogen storage, up to 7 percent. The implied weight of a system would be 424 pounds, with a cruising range of 300 miles and a volume of 57 gallons.

The Daimler-Benz hydrogen vehicle program has used hydride storage systems since the 1970s. From 1984 to 1988, their Berlin fleet test vehicles were all equipped with metal hydride storage systems. The hydride container consisted of nineteen high-grade steel tubes filled with an alloy of titanium, vanadium, and manganese. A liquid media flowed through the storage unit between the tubes and outer castings to heat the hydride and release the hydrogen. This system stored a maximum of 33 pounds of hydrogen and weighed 313 pounds.

DISTRIBUTION

If a hydrogen economy is to develop to its fullest, a distributing system must be in place to deliver the product to the consumer. Here, the infrastructure of fossil fuels plays a role. During the transition to a sustainable solar hydrogen economy, hydrogen must be acquired from fossil fuels. Hydrogen will be processed from natural gas, which has in place a distribution system throughout the United States. Fortunately, hydrogen gas can be delivered through this established infrastructure.

Pipelines

The natural gas delivery system includes 1.3 million miles of pipeline. During the initial stages of the transition to hydrogen, these pipelines could be used to deliver hydrogen. For a full-scale hydrogen economy, this system would not be adequate. Distribution techniques needed for hydrogen are similar to natural gas.

Natural gas is produced at the well head or manufacturing plant and piped to city gate stations, whose function is to meter the natural gas and reduce the gas pressure of the transmission pipeline to that of the distribution system. Before the gas leaves a city gas gate, a small controlled amount of substance with a strong penetrating odor is injected into the gas. The odorant warns people of the presence of unburned gas before it accumulates in hazardous quantities.

From a city gate station, the gas is piped to the remainder of the distribution system in a supply main. The supply main feeds the trunk mains that go to feeder mains, and finally, to distributor mains that supply the customers. Fuel lines on the premises of residential customers normally operate at low pressure.

The operating conditions for a distribution line that delivers hydrogen are determined by the equivalent fuel value of hydrogen to that of existing natural gas. A higher volume of hydrogen gas would need to be delivered to equal natural gas by a factor of between 2.9 and 3.4. The higher heating value of hydrogen is 325 Btu per standard cubic foot, compared to 1,060 Btu per standard cubic foot for natural gas. Because of the increased volume, volumetric flow meters used for measuring gas delivery may be undersized for hydrogen usage. The greater volume of hydrogen that must be pumped through the pipeline compared with natural gas would require the installation of pumps and compressors that operate at higher pressure than the present ones. Most of the equipment would have to be modified to be compatible with hydrogen.

The sulfur compounds used to odorize natural gas are compatible with hydrogen. Because the hydrogen flame is invisible, the simultaneous addition of an illuminant would be worthwhile, facilitating the use of hydrogen in open flame appliances, such as stoves.

Because hydrogen is such a small molecule, the rate of leakage through orifices, cracks, corrosion pinholes, or leaky seals is about 2.5 to 3 times greater for hydrogen than for natural gas,

The permeability of plastic pipes to hydrogen compared with natural gas is another concern. The ratio is 7:1 for polythene and 88:1 for AB1. However, in comparison with typical natural gas distribution systems, the hydrogen loss by permeation through plastic appears insignificant.

Embrittlement, the deterioration of a metal pipeline due to a chemical reaction with hydrogen, is a concern. Some examples of embrittlement include decarburization (losing carbon) of steel and the formation of irregular hydrides. The exact cause of embrittlement is not understood. Environmental embrittlement is the degradation of mechanized properties caused by absorption of hydrogen at the surface of a metal. Some experts believe the metallurgical effect would not occur in distribution using natural gas lines. Most of the pipes are made from plastic. However, others believe that embrittlement could be a problem necessitating replacing the pipes. In fact, they suggest that the development of a dedicated hydrogen pipeline would prove cost-effective in the long run.

Working hydrogen pipelines have been proven to be feasible over time and distance. The oldest hydrogen pipeline was built in 1938 in the Ruhr Valley of Germany and measures 130 miles long. Between Belgium and northern France lies the world's longest pipeline, a 250-mile pipe operated by L'Air Liquide. The United States has 450 miles of hydrogen pipeline, with Praxon Company operating three lines in Texas, Indiana, and New Jersey that total 160 miles. Two hydrogen pipelines measuring 100 miles are operated by Air Products and Chemicals, Inc. in Texas and Louisiana.

When the transition to hydrogen has progressed, hydrogen can be piped to a refueling station for compression, much like natural gas is compressed at refueling stations. Eventually hydrogen will be produced on-site, perhaps from electrolysis.

These small production units will be developed when they become economically competitive.

Other Distribution Methods

Trucks can also transport hydrogen as a high-pressure gas in cylinders at pressures ranging from 2,000 to 6,000 psi. The cylinders are large, high strength, high pressure steel tubes that have been manifolded and mounted on a truck trailer. Highly efficient, this system can haul small volumes of hydrogen gas. A steel cylinder weighing 40 to 60 pounds will hold 1 kilogram of hydrogen at 2,000 psi and 2.5 kilograms of hydrogen at 6,000 psi. Because of the large container weight, hydrogen comprises only 2 to 4 percent of the cargo weight.

Liquid hydrogen is transported in batch quantities over public roadways from the producer to user by truck. These semi-trailers consist of a cryogenic liquid hydrogen tank that has concentrical inner and outer tanks separated by an evacuated space filled with perlite insulation. Liquid hydrogen is carried in the inner tank and is effectively insulated from the environment by the annular evacuated space. Transportation costs of liquid hydrogen are much higher than for gaseous transportation.

Railroad tank cars are also used to transport liquid hydrogen, and capacity ranges from 9,500 to 28,000 gallons. The tank cars are equipped with insulation layers that consist of vacuum jackets and a multilayer radiation shielding to keep evaporation losses to below 0.5 percent per day.

Today, dedicated hydrogen ships do not exist. NASA uses three liquid hydrogen barges for the space program, with a capacity of 250,000 gallons. The Euro Quebec Hydro Hydrogen Pilot Project is currently designing liquid hydrogen tanks for their ships. An innovative design seeks to prevent hydrogen fuel boil-off for at least 50 days. Presently, 100 tanker ships for liquid natural gas operate between Europe, North America, South America, and the Pacific Rim.

Once the tanker ships arrive at the port, the hydrogen must be transferred to a storage tank. These transfers could cause a 20 percent energy loss. Demand will have to rise before justifying a dedicated tanker.

REFUELING

Consumers have been spoiled by the ease and quickness of refueling their gasoline vehicles. For consumers to purchase hydrogen vehicles, they would prefer not to be inconvenienced. During the hydrogen transition, refueling stations will not be conveniently located compared with present-day gasoline stations. Initially, there would only be a few. When the infrastructure develops, present-day gas stations will be used to refuel hydrogen vehicles. Tanker trucks can deliver the hydrogen to the gas stations or natural gas pipelines can be used. The hydrogen will be compressed into storage tanks. As demand grows, hydrogen will be produced locally, most probably by electrolysis.

Fleet vehicles of government agencies and commercial fleets can help initiate the hydrogen transition. Each facility would have a refueling station to service the fleet vehicles, whether a transit bus fleet, support vehicles at an airport, or a fleet of municipal cars.

Refueling a hydrogen car requires a different technology. For metal hydride vehicles, a practical cold reservoir, a culinary water system, for cooling the vehicle hydride tanks is required during refueling. When hydrogen is supplied at sufficient pressure, this is the only heat reservoir required for refueling. The tanks must be kept cool, because heat releases the hydrogen from the metal hydride.

Hydride storage systems require a heat source at high enough temperature to build sufficient pressure for flow into the vehicle tank. The use of hot and cold reservoirs are the most attractive way to provide for heat transfer between the vehicle and station hydride units. The best cold-water source, the culinary system, provides flow through a hot-water system located in the station building.

The metal hydride vehicle must undergo recharge from the station hydride that discharges at elevated pressure. An alloy releases heat as it hydrides, and the resultant heat must be removed for the reaction to continue to maximum value.

Recharging metal hydrides can take some time. During the recharge of the Billings hydrogen bus, within five minutes, 40 percent of a three-hour recharge was complete. After 30 minutes, 78 percent was complete, and 86 percent was complete in one hour. It is advantageous to pull into the refueling station with a cold hydride. The remainder of the refuel period is limited to heat transfer. During the recharge, after the initial rapid flow, heat must be removed from the hydride.

A hydrogen refueling system was centrally located at the Berlin gasworks for the Daimler-Benz vehicles and their metal hydride storage system. Hydrogen was obtained from the city gas containers that contained approximately 50 percent hydrogen by volume that used a pressure surging adsorber for refueling. The tanks were continuously checked to ensure the hydrogen was greater than 99.9995 percent pure. Under a pressure of 750 psi, hydrogen was administered by quick coupling. Within ten minutes, the vehicle was refueled to 80 percent capacity.

At filling stations for liquid hydrogen, the transfer of deep-cold liquids requires special, insulated transfer lines. The cryogenic coupling is the critical part of liquid hydrogen refueling. Couplings have to be deep inside one another and surrounded by thermal insulation. This equipment can be quite costly. Only experienced and qualified technicians can handle the equipment. In Germany, BMW is experimenting with robotics to do the refueling. If one is not careful, ice can develop during the dismounting and the coupling will have to be heated.

The Linde Company developed a novel hermetic coupling ten years ago to avoid this ice problem. The first public liquid hydrogen filling station at the Munich Airport has a hermetic, clean-break coupling. To refill a liquid hydrogen vehicle tank with 25 gallons of hydrogen takes less than two minutes.

By 2005, BMW expects to have one liquid hydrogen filling station close to every BMW subsidiary. They also expect that each major European city will have at least one hydrogen refueling station.

Another type of liquid hydrogen refueling station was built on the Bavarian Solar Hydrogen Company site. Liquid hydrogen was "rained in" from above. On its way through the tank, the hydrogen, already evaporated, condensed into droplets, preventing any loss of hydrogen. This system not only reduced the fuel loss to zero, but made it possible to reduce the filling time from the original one hour to less than three minutes.

Although the hydrogen infrastructure seems to be workable, hydrogen storage seems to be the biggest stumbling block. Consumers are used to convenience, and at this stage, they are not willing to give up their trunk space for hydrogen storage tanks and a short driving range. As we will see, the hydrogen fuel cell may be the answer to storage, driving range, and a bridge from fossil fuels to the hydrogen future.

Chapter Four

THE FUEL CELL

The Ultimate Power System?

As we have seen, hydrogen seems to be the ideal nonpolluting fuel that has the potential for saving our environment. The major drawback is storage. Gaseous hydrogen requires considerable volume for storage tanks. Not many consumers want to sacrifice trunk space for hydrogen storage. If they do not sacrifice trunk space, they sacrifice driving range. Liquid hydrogen does not require as much storage volume, but it must be kept at -453 degrees Fahrenheit. This is not practical on a large scale, but it can be done. Metal hydrides used for storing hydrogen in an internal combustion vehicle add too much weight and require much more storage space compared with gasoline. Storage problems seem to be the "albatross" that is keeping hydrogen from competing with fossil fuels. However, all is not lost!

Scientists are becoming very excited about the fuel cell that can power an electric vehicle. Many of these scientists believe fuel cell technology will be the power delivery system of the future. Great strides in fuel cell technology have occurred within the past five years. Fuel cells can use hydrogen, our ideal fuel. They can provide the power produced in power plants; they can be used as the power source for vehicles; and today, they are being produced for home use, allowing people to get off the power grid. This all sounds impressive, but just what is a fuel cell?

The Fuel Cell

Fuels cells are conceptually very simple and very efficient; they have modular capabilities; they are environmentally friendly; and they are very flexible, allowing for a variety of applications. Fuel cells produce their energy through a direct and reversible chemical process that allows fuel cells to operate at an efficiency two to three times greater than that of a conventional internal combustion engine.

Essentially, a fuel cell is a galvanic cell, or battery, that converts chemical energy into electrical energy through an electrochemical process. Similar to a battery, the fuel cell is a device in which electrons are forced to flow through an external pathway (a wire) rather than directly between reactants. At separate electrodes, the cell operates by taking in fuel and an oxidant, with the latter usually being pure oxygen or air. Excess chemical energy is then converted into direct current, DC electricity. The fuel source is normally hydrogen, or a hydrogen carrier such as natural gas or methanol. When operated with hydrogen and oxygen, water and heat are the only byproducts.

A generic fuel cell consists of two porous carbon electrodes, where energy conversion takes place, and a solid or liquid electrolyte, which conducts ions to form a closed circuit. A wire connects the two electrodes. Hydrogen molecules are introduced on the anode side, sometimes called the fuel electrode. The hydrogen molecules give up their electrons, forming positively charged hydrogen ions ($H_2 = 2H+ + 2e-$). The cathode attracts these electrons with the oxidizing agent, oxygen, resulting in negatively charged oxygen ions. The electrons are prevented from going directly to the cathode by a membrane, so they travel through the wire. Electrical energy is created as the electrons move through the external circuit connecting the cathode and anode, and this electricity becomes the power source. The medium through which the ions migrate from one electrode to another is the electrolyte. It provides a pathway for the hydrogen ions (protons, H+) to pass from the anode to the cathode side of the cell.

The fuel cell also contains a membrane catalyst interface. Here the hydrogen ions combine catalytically with oxygen, which has diffused to this interface through the porous cathode. The electrons that split off from the hydrogen molecule combine with oxygen and the hydrogen ions to form water ($4H+ + O_2 + 4e- = 2H_2O$). The overall reaction is $2H_2 + O_2 = 2H_2O$. When the hydrogen hits the electrode and the electrons are stripped off, the electrons can carry a current through the wire to the other electrode. The amount of energy created in the current is almost the same as if

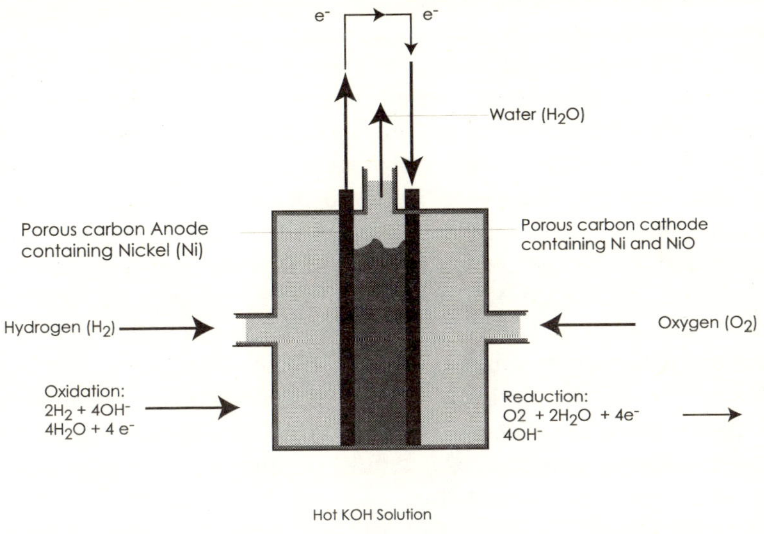

A FUEL CELL

two chemicals were combined explosively, but here the energy is available at a rate that enables it to do useful work.

A fuel cell operates at higher efficiency than a heat engine because chemical energy is directly converted to electrical energy. A heat engine converts chemical energy to mechanical energy with heat as an intermediary. Several mechanisms reduce the optimum efficiency of fuel cells. For example, auxiliary energy is required for power conversion (converting direct current to alternating current) and for fuel processing. The ideal efficiency for a hydrogen/oxygen fuel cell is 83 percent if the product is water, and 94 percent if the product is water vapor. Operating fuel cells achieve an efficiency of around 40 to 45 percent. The efficiency of a heat engine can be 30 percent with an overall efficiency of 25 percent or less.

History

Sir William R. Grove, a lawyer and knighted judge, developed the world's first fuel cell in 1839. He used hydrogen as the fuel, oxygen as the oxidizing agent, and dilute sulfuric acid as the electrolyte. A. C. Becqueral later developed a carbon-burning cell that employed a fused-cell electrolyte featuring a carbon and a platinum electrode immersed in a nitrate. At the turn of the twentieth century, electricity was gaining in popularity and coal plant generation was only 2.6 percent efficient. This led to the drive for developing an efficient source of electricity.

Dr. W. W. Jacques then developed a fuel cell for power generation that was 80 percent efficient and used carbon as a fuel.

Francis W. Bacon of Cambridge University developed the first major cell of the twentieth century. He researched the hydrogen/oxygen alkaline fuel cell using nickel electrodes, thinking that the expensive platinum catalyst systems would prevent mass commercialization. Bacon believed that when nickel was heated to 205 degrees centigrade, it would eliminate the need for any other catalyst. Bacon was the first to introduce porous gas diffusion electrodes to increase the contact between the gas and electrodes. By 1959, he had demonstrated a 5-kilowatt system capable of powering a welding machine, a circular saw, and a two-ton forklift truck.

Bacon's technology laid the foundation for fuel cell systems in the Gemini and Apollo space flights. NASA realized the potential of fuel cells for space application and developed them even further. General Electric and United Technologies Corporation became the leaders in fuel cell technology after receiving contracts from NASA.

In 1973, the energy crisis caused by the Arab oil embargo stimulated interest in fuel cell-powered vehicles. Great strides have been made in recent years, and the research continues.

Types of Fuel Cells

There are two categories of fuel cells, low temperature and high temperature. Low-temperature fuel cells have an alkaline or solid-polymer electrolyte. High temperature fuel-cells include phosphoric acid fuel cells (PAFC), solid oxide fuel cells (SOFC), and molten carbonate fuel cells (MCFC). Today, the two most common fuel cells are the alkaline fuel cells that have been used extensively in the U.S. space program and the phosphoric acid fuel cell. The alkaline, solid-polymer fuel cell is being researched extensively as the most promising for vehicular usage. High-temperature fuel cells, SOFC and MCFC, are being developed for stationary utility applications.

One of the most developed fuel cell technologies is the alkaline fuel cell. Using less expensive metals, such as nickel, for the electrocatalyst, the cells exhibit good performance. These fuel cells are efficient, reliable, easy to operate, and very important for vehicles. They also have low weight and volume.

The major drawback is that the electrolyte of alkaline fuel cells is extremely intolerant to carbon dioxide. Vehicle systems would probably have to be supplied either with pure oxygen or with air scrubbed of carbon dioxide. Even the small levels of carbon dioxide in the air (350 ppm) can carbonate the electrolytes to form deposits on the porous electrodes.

The electrolytes for alkaline fuel cells can be mobile or immobile, based on a porous matrix such as asbestos. A mobile electrolyte continuously circulates between the electrodes. Water and heat in an immobile electrolyte do not continuously circulate and have to be removed by other means. Hydrogen gas circulation removes water from the cell when it evaporates and condenses. A circulating coolant removes heat waste.

A polymer fuel cell or proton exchange membrane (PEM) fuel cell is the most attractive application for vehicles. General Electric first developed it for NASA in 1961, and it has a high energy density compared with batteries. The PEM fuel cell has no corrosive liquid electrolytes; it has a simple stack design and is very rugged. These fuel cells were used in seven Gemini missions and lasted approximately 500 hours.

By the 1960s, General Electric adapted Dupont's Nafion for use in the solid polymer fuel cell (SPFC). This teflon-like material's operating lifetime exceeded 57,000 hours. For some reason, NASA decided on the alkaline fuel cell for the Apollo program, and the solid polymer fuel cell was put on hold until 1984. Then the Canadian Department of National Defense determined that the SPFC could be used for a compact power supplier. Ballard technologies of Canada developed the SPFC further. In the United States, Los Alamos National Laboratories, Texas A&M University,

Hamilton Standard, and International Fuel Cells have researched the SPFC.

The SPFC consists of two porous electrodes bonded to either side of a thin sheet of an ion-bonding polymer in a solid-polymer electrolyte. The backs of the electrodes contact plates that contain channels through which hydrogen and oxygen are supplied. Electrical contact to the electrodes occurs through these plates. The waste byproduct is produced at the cathode and removed, and a circulating coolant removes waste heat.

The polymer electrolyte of a SPFC is a plastic-like membrane. These membranes are composed of perfluoro sulfonic acid, teflon-like fluorocarbons, called proton exchange membranes (PEM), about four thousandths of an inch thick. They are excellent conductors of hydrogen ions, but prevent electrons from passing through. Until recently, Nafion 117 was the most used membrane in SPFCs.

In 1987, Ballard technologies demonstrated that another Dow membrane could generate a larger current and power density in fuel cells. The electrodes are porous, allowing for effective gas diffusion. They also pass the generated current to the cell. A catalyst layer, composed of platinum, is added to the electrodes to increase their effectiveness. Recent studies at Los Alamos and Texas A&M have shown electrodes loaded with low levels of platinum are more effective than those loaded with higher levels. This is good news, since platinum drives up the cost of fuel cells.

The high-temperature fuel cells, PAFC, MCFC, and SOFC can tolerate carbon dioxide, and they operate on fuel mixtures produced from coal gasification, reformed alcohols, and natural gas. SOFC and MCFC have been developed for stationary utility application, offering an environmentally compatible way to produce electricity at competitive costs. PAFCs can be used for both stationary sources and for vehicle applications.

Phosphoric acid, used in PAFCs, is the only common acid capable of functioning in a fuel cell, at temperatures much higher than water's boiling point. It has low volatility and is stable. PAFCs are sensitive to carbon monoxide, which must be reduced to about 1 to 2 percent.

MCFCs (molten carbon fuel cells) operate at temperatures of

1,200 degrees Fahrenheit. These cells are highly efficient with extremely low emissions. They operate on natural gas or gas produced from the gasification of coal, the cleanest way to use coal. Water and carbon dioxide are its only emissions. The major problems associated with MCFCs are corrosion and sintering of the electrode material that causes the electrode to dehydrate. Platinum metals are one of the few materials resistant to molten carbonate electrolysis.

Solid oxide fuel cells (SOFCs) use a solid oxide material, normally zirconia combined with other rare earth materials, as an electrolyte. When the electrolyte is heated to 1,000 degrees centigrade, it conducts ions at a sufficient rate to act as a solid state electrolyte. Westinghouse Electric Company is the primary manufacturer of this cell, and the company operates several pilot plants in Japan and the United States. SOFCs can oxidize any gaseous fuel, and their chief advantage is operating by transport of oxide ions rather than fuel-derived ions. SOFCs have good tolerance for fuels with high levels of impurities, which makes it feasible to operate with heavier fuel such as coal gas. SOFCs are compatible for industrial use.

FUEL CELLS FOR VEHICLES

Many scientists are concluding that the hydrogen fuel cell will be the future power source for automobiles. The cell could combine the best features of battery-powered electric vehicles: zero emissions, quiet operation, and long life. It will also provide the benefits of the internal combustion engine: a long driving range and fast refueling time. The solid polymer fuel cell (SPFC) appears best suited for vehicle application. It provides room-temperature startup and eliminates many corrosion problems. The solid electrolyte facilitates the sealing and safety of the fuel cell stack. SPFCs can tolerate carbon dioxide in the fuel systems, but not carbon monoxide.

An important requirement of a fuel cell system is to provide sufficient power. Auxiliary batteries have been used to supplement the fuel cell during periods of peak operation, such as acceleration.

Cost has always been a major concern regarding fuel cells.

Cost reduction is a great challenge with regard to proton-conducting membrane systems and platinum-catalyzed electrodes. Perhaps researchers at Los Alamos National Laboratories have helped solve part of this problem with their thin-film technology. They estimate that an 80-kilowatt-power passenger car would require less than $500 worth of platinum. The polymer membrane cost is a bigger issue. Dupont's Nafion and Dow's state-of-the-art membrane cost about $60 to $200 per kilowatt. An 80-kilowatt system would cost between $4,800 and $16,000, too much for most consumers. However, if the membranes were marketed and the demand increased, some studies suggest that the costs per car could drop to between $300 and $400 a car, a cost most people could afford. If fuel cells could be successfully developed on a life-cycle basis, their costs could become competitive with gasoline internal combustion engine vehicles.

A complete fuel cell system requires various components, including: (1) the fuel cell stack itself, (2) a storage system for fuel (either hydrogen or methanol), (3) an air compressor to pressurize the gases, (4) a cooling system, (5) a water management system to keep the fuel cells saturated and to remove end-product waters, (6) an electric motor and controller, and (7) a high-powered battery to supply power at peak periods.

The electrical storage system in fuel cell electric vehicles consists of a battery bank that can be charged by the fuel cells, or by energy lost in braking, using a "regenerative braking" system. The electric storage system would be larger than a battery used in conventional vehicles, but it would be a smaller storage system than batteries required for electric vehicles.

Hydrogen could be stored onboard in the form of a hydrocarbon and then processed to a hydrogen-rich gas using a reformer. Methanol is the best hydrocarbon to use in the reforming process. Problems of storing pure hydrogen onboard would be avoided. Metal hydride storage cylinders could also be used.

Until recently, the fuel cell car would need to be larger and heavier. A fuel cell drive line would be bulkier and heavier compared to an internal combustion engine of the same power by a factor of three. The power train of a fuel cell car would cost two to three times more than an efficient internal combustion engine.

Other Fuel Cell Developments

Leading fuel cell manufacturers and utility companies in the United States, Japan, and Europe are actively trying to introduce fuel cells into the commercial market. Utility companies in Colorado and the Pacific Northwest are marketing fuel cells for the home, allowing the consumer to become independent of the public utility grid.

Japan is a major pioneer in fuel cell technology. The Japanese Ministry of International Trade and Industry established a target of 2,250 megawatts of fuel cell capacity by the year 2000. The Tokyo Electric Power Company developed a single operational fuel cell stack that is generating 11 megawatts of power. Located in Ichihara City on Tokyo Bay, the plant uses phosphoric acid fuel cells that supply power to approximately 400 homes. A joint venture between International Fuel Cells (IFC) and the Toshiba Corporation enabled the plant to be constructed.

Fuji Electric, a leader in the development of phosphoric acid fuel cells, has committed considerable investment toward commercialization of PAFCs for areas of on-site cogeneration, with systems ranging from 50 to 500 kilowatts for power plants in the megawatt class and for vehicles requiring less than 100 kilowatts. The company's factory has a manufacturing capacity of up to 15 megawatts.

Another major player in fuel cell technology is the International Fuel Cell Corporation and its subsidiary, the ONSI Corporation. The company has been the largest producer and installer of fuel cells in the world. They built the 11-megawatt facility in Japan and a number of commercial 200-kilowatt PAFC plants in the United States. In 1994, they had orders for sixty-nine such plants.

Molten carbonate fuel cell (MCFC) technology is being developed in the United States by the Energy Research Corporation of Danbury, Connecticut. They are building a 2-megawatt MCFC plant in Santa Clara, California. Southern California Utilities is funding this $46-million venture.

M-C Power Corporation is building a 250-kilowatt MCFC in the area served by San Diego Gas and Electric Company. A 250-kilowatt MCFC plant will be operated at the Science and Technology Center of UNOCAL in Brea, California.

Westinghouse Electric Corporation is a significant developer of solid oxide fuel cells. The company has two 25-kilowatt units operating at Rokka Island Test Center for Advanced Energy Systems near Kobe, Japan.

H-Power Corporation was the primary contractor for a Department of Energy project to develop three fuel cell passenger buses. They were powered by phosphoric acid fuel cells in conjunction with nickel-cadmium batteries. The first bus was unveiled in April,1994.

In the mid-1990s, General Motors participated in a multiphase Department of Energy (DOE) program to demonstrate the feasibility of proton exchange membrane (PEM) fuel cells and a methanol reformer for transportation application. The DOE program teamed General Motors with fuel cell scientists to transfer technology to the industrial arena. In 1967, GM had experimented with a fuel cell/battery-powered van that used hydrogen and oxygen. However, the storage tanks consumed too much space.

Ford Motor Company and DaimlerChrysler received $29 million from the DOE for fuel cell development. Ford's research focused on reducing the cost, size, and weight of fuel cells. They also investigated hydrogen production, distribution, storage, and dispensing. The DOE sought to have prototype cars produced by 2004, a goal that these companies may reach sooner.

Engineers at Ford Motor Company replaced the conventional drive train (consisting of the engine, automatic transmission, and exhaust system) of a Mercury Sable with a fuel cell system and a 5,000-psi compressed hydrogen tank. The design engineers showed that an advanced fuel cell system, including compressed gas tanks holding 3.6 gallons of hydrogen, could be packaged into a five-passenger car without intruding into the trunk space. The hydrogen fuel cell car had the same acceleration as the conventional Sable and could travel 285 miles. A prototype vehicle is being developed with this design that has a range of 380 miles, with a gas equivalent of 110 miles per gallon.

DaimlerChrysler researched the PEM fuel cell's use with gasoline, which could advance fuel cell technology by ten years. The prototype improved fuel efficiency by 50 percent and was

90 percent cleaner than gasoline engines. Essentially, the vehicle produced electricity onboard through a five-step process: First, it converted gasoline to a gas and then combined it with air to produce hydrogen and carbon monoxide, followed by converting carbon monoxide to carbon dioxide that was converted to water. Finally, hydrogen gas was combined with air to produce electricity via the fuel cell. The prototype vehicle had virtually no pollution, emitting only water vapor.

Research at the Center for Electrochemical Systems and Hydrogen Research at Texas A&M University has investigated PEM fuel cell projects since 1988. Their research showed how to reduce platinum loadings on electrodes by a factor of 100.

The Canadian company Ballard Power Systems has built and is testing a 21-passenger transit bus entirely powered by a PEM fuel cell. In 1995, General Motors contracted with Ballard Power Systems to produce fuel cells for an automotive engine and for integration into a hybrid fuel cell engine.

The U.S. company Energy Partners has demonstrated a battery/fuel cell-powered car called the "Green Car." Both Mercedes and Mazda have demonstrated a fuel cell-powered van and a fuel cell golf cart. Fuel cell buses are being developed as part of the Euro Quebec Hydro Hydrogen Pilot Project.

Another use of fuel cells is to increase the range of electric cars. With the aluminum-air fuel cell, Alupower Canada, Limited, equipped a Chrysler minivan with the company's Alupower Range Extender. This fuel cell, which weighs 860 pounds, extended the range of an electric van from 47 miles on lead acid batteries to 186 miles. The technology is based on a refuelable aluminum fuel cell. Passenger space was not affected in this prototype.

Another experimental fuel cell battery is Sanyo's Mirai-1. It uses the energy-related technologies of solar cells, nickel-cadmium batteries, and phosphoric acid fuel cells. The solar cells and/or the fuel cells charge the batteries, which drive a direct-current motor. On a fuel charge, it can run for two hours at a top speed of 62 mph. The Sanyo CEO does not believe that fuel cells will completely displace batteries because fuel cells take several minutes to come to full power, and it is better to run the vehicle at a constant output. Sanyo envisions fuel cells paired with batteries, to provide needed power for typical driving. The potential advantage of fuel cells is fewer batteries and the ability to recharge them on the go. The present energy output of the fuel cell is 3.8 watts per pound, less than one-third that of a lead acid battery.

In Germany, a joint venture between two utility companies in Hamburg operates two phosphoric acid fuel cell facilities. One of the fuel cells was fueled by natural gas and put on line in 1995. The hydrogen system was put on line in 1997. Each fuel cell contributes 200 kilowatts of electricity fed into the 10-kilovolt grid system of one utility and a 220-kilowatt thermal power system of another. The latter was used in combination with an existing heating system to heat a 680-unit apartment building complex at the site. The heat supply capacity of both fuel cells provides heat for 40 apartments and electricity for 100 apartments.

In the spring of 2000, Ballard Power Systems provided a second 250-kilowatt unit cell to a consortium of German utilities in which a natural gas cogeneration plant provided heat and electricity in Berlin.

DaimlerChrysler, Ford, and Ballard Power Systems are working on a high-profile joint project to commercialize a fuel cell vehicle by 2004. Toyota announced in 2000 that it plans to market a fuel cell vehicle in 2003. General Motors and Jet Propulsion Laboratories are also working on a fuel cell project. At this time, the total worldwide investment in fuel cell research and development for light-duty vehicles is in the multibillion-dollar range. Fuel cell vehicles that use hydrogen will produce no pollution, and they will meet the California standards for zero emissions in 2003.

Other fuel cell vehicles that carry reformers can derive their hydrogen from methanol, ethanol, gasoline, and natural gas. These fuels can be distributed within the existing infrastructures at reasonable cost. These vehicles produce near-zero emissions.

Four companies, Plug Power in New York, Avista Laboratories in Washington, Northwest Power Systems in Oregon, and Delta Montrose Electric in Colorado have fuel cell units that provide electricity to homes. A fuel cell of about 10 kilowatts can meet the need of a large four- or five-bedroom home or a small commercial establishment. An automobile requires a 50-kilowatt cell to provide power acceleration.

Epilogue

Fuel cells represent a link between the fossil fuel economy and a future economy based on hydrogen/electricity. They are the most promising technology for hydrogen because of their high efficiency and low maintenance costs. During the transition to a hydrogen economy, fuel cells can operate on fossil fuels through a reforming process. During this stage, fuel cells can be supplied with natural gas or directly with hydrogen. The California Clean Air Act provided the greatest impetus for automobile manufacturers to look seriously at fuel cells. Nearly all automobile manufacturers are making an effort to commercialize fuel cell vehicles between 2003 and 2005. Some will operate on pure hydrogen and others on methanol or natural gas. The necessity of hydrogen refueling stations has now become a reality. The greatest issue for the manufacturers is to choose a fuel type for their vehicle, that is, from among hydrogen or several kinds of hydrocarbon-based fuels.

The Japanese founded the WE-NET Hydrogen Refueling Station Project in 1999, with the supposition that fuel cell vehicles with metal hydride storage tanks vehicles would refuel at WE-NET stations. They chose hydride storage due to regulations regarding high-pressure gas, which a metal hydride system could pass. The Hydrogen Refueling Station Project developed two prototype stations with different types of hydrogen generators. One station uses water electrolysis for hydrogen production and the other uses natural gas reformation. Both the water electrolysis and natural gas reforming stations are being constructed in Japan. The two stations will generate hydrogen at 1,060 cubic feet per hour.

Metal hydride was selected as the best hydrogen storage system in both types of hydrogen refueling stations. Its advantage is its capability of storing the largest quantity of hydrogen for practical use in fuel cell vehicles. The LA-Ni-Mn metal alloy was selected as the hydrogen adsorbent, as its adsorption/desorption temperature is relatively low. Because of nonconformity in national high pressure gas regulations, metal hydrides may have the greatest hydrogen storage potential for fuel cell vehicles.

In California, the number of hydrogen vehicles may reach 30 million by the year 2010. At that time, fuel demand will reach 22.7 billion gallons of hydrogen. On an equivalent energy basis, 8.8 trillion cubic feet of hydrogen would be required if all vehicles were filled with hydrogen at the same automobile efficiency. Scientists conclude that if hydrogen-fueled vehicle manufacturing is commercially available by 2010, the electric power to produce hydrogen would not be a significant deterrent during the start-up phase through the year 2025. By the year 2050, the zero-emission hydrogen vehicle replacement would be reached. During this period, the need for additional electric power facilities will become acute.

As the cost of proton exchange membrane (PEM) fuel cells declines, hydrogen PEM fuel cells will eventually replace the traditional generators fueled by diesel and gasoline. These generators are quieter, more reliable, have no noxious emissions, and are ideal for commercial, residential, and industrial backup power. As mentioned earlier, some utility companies have already begun to market their fuel cell generators for home use.

These hydrogen fuel cells will be able to provide digital quality power for telecommunication stations. The metal hydride storage modules can supply hydrogen to the PEM fuel cell unit to generate electricity for night use. In the case of cogenerated hydrogen, hydrogen will be transported to the central station or the village. Residents will go to the station to refill or exchange metal hydride storage modules to keep their hydrogen fuel cell generator running.

The hydrogen fuel cell may well be the power source of the future. These cells will turn hydrogen directly into electricity at very high efficiency. This efficiency is two to three times higher than achieved with current internal combustion engines. Hydrogen fuel cell vehicles have zero emissions and the reactive end product is pure water. During the hydrogen production process, when water is electrolyzed, oxygen is produced from water. The oxygen produced can alleviate many types of pollution unrelated to fossil fuels. It can enhance the treatment of sewage, purify organic contaminants in lakes and rivers, and convert ozone to replace chlorine in drinking water. Fuel cells may ultimately generate all electricity locally, eliminating overhead transmission lines. Fuel cells also would be more reliable than conventional power plants. The future is bright for the hydrogen fuel cell.

Chapter Five

HYDROGEN SAFETY

Overcoming the Hindenburg Syndrome

In 1937, the German hydrogen airship Hindenburg caught fire as it landed in Lakehurst, New Jersey. Cameras caught the spectacular flames, and motion picture theaters around the world replayed the spectacle. This created a consciousness among the public that hydrogen was unsafe and highly unstable. The attitude persists today and is called the "Hindenburg syndrome." Experts now understand that not one person died from the hydrogen fire, but almost all of the fatalities occurred as a result of people jumping to the ground. A recent study by a NASA investigator has shown that hydrogen did not contribute to the severity of the disaster. Within one minute of the time the fire began, all the hydrogen gas was probably consumed. The fire, which burned for over an hour, was fed by the diesel oil carried for fueling the engines. As we will see, hydrogen is as safe as or safer than gasoline.

The flammable properties of hydrogen were known in the 1780s when Pilatre de Rozier chose to demonstrate these properties by breathing in hydrogen and then setting fire to the gas as he exhaled it through a glass tube. On one occasion, he repeated the experiment with a mixture of hydrogen and air. The resultant explosion left de Rozier thinking he had blown out all of his teeth. Undaunted by this experience, in 1784, de Rozier built a balloon that took advantage of the buoyancy of both the hot air and hydrogen. Ignoring the warnings of having an open fire underneath a balloon filled with hydrogen, de Rozier believed the hydrogen would rise away from the flame below. In 1785, a flash of flame ignited the balloon and de Rozier crashed to the ground, giving him the distinction of being the first astronaut to die. The point being, hydrogen can be very dangerous without proper precautions.

Safety Concerns for Hydrogen

Hydrogen has some properties that suggest it may be a greater hazard than other combustible fuels such as gasoline or natural gas. Hydrogen has low ignition energy, a high flame velocity, and a small quenching distance.

The high ignition temperature gives hydrogen an advantage over other fuels. For example, a lighted cigarette will not ignite hydrogen in air. However, if hydrogen accumulates, a cigarette lighter will ignite a hydrogen/air mixture.

Hydrogen has a wide range of combustibility and detonatability. The safety of a situation depends on maintaining the atmospheric hydrogen gas level below the lower explosion limit. The lower explosive limit of hydrogen is similar to methane, which is 4 to 5 percent. This means that an accumulation of hydrogen exceeding these limits will cause an explosion if ignited. Gasoline vapor has a lower explosion limit, meaning that hydrogen is not necessarily more dangerous to handle compared with gasoline. Experts conclude that the wide range of combustibility is not necessarily a detriment to safety in handling hydrogen.

Hydrogen has a high diffusivity, meaning it will mix quickly with air. In open air, hydrogen would be safer than slower diffusing gases. However, in confined spaces, hydrogen mixes quickly with air to reach all parts of the space, thereby making it less safe.

Hydrogen has a faster dissipation rate compared with other gases. This means that for a given hydrogen release, at a given time, only a small portion of hydrogen may exist in concentrations within the combustible limits. In other words, because hydrogen is so light, it leaves the scene quickly. For this reason, the total energy release from a hydrogen explosion is estimated at only a few percent of the total energy available.

Because hydrogen has a low flame emissitivity, a hydrogen flame will radiate less energy, reducing the likelihood of causing burns to personnel at a given distance from the flame. However, this low emissitivity makes the flame invisible and creates an additional hazard if someone inadvertently enters a hydrogen flame.

Hydrogen has a very high electrical resistance, meaning a reduced likelihood of electric charge accumulation in flowing liquid hydrogen.

A large volume ratio exists between liquid hydrogen and room temperature hydrogen, as happens with any cryogen. To maintain the liquid hydrogen in a closed container requires high pressure. The low boiling point implies the existence of a large thermal driving force attempting to pull heat from the surroundings into the liquid hydrogen. Because no perfect insulation exists, this heat can only be minimized. The low-volume latent heat of vaporization means that larger quantities of stored liquid hydrogen will be evaporated per unit of energy input. A small percentage of stored liquid hydrogen evaporates each day. These properties of hydrogen make it a potentially dangerous and highly explosive fuel. However, other fuels such as gasoline and natural gas have similar dangers.

Because of strict adherence to regulations and procedures, hydrogen has an excellent safety record. Since the 1940s, hydrogen has been shipped through extensive pipeline networks in Germany. Mercedes-Benz and BMW have operated a fleet of hydrogen vehicles for a number of years without a mishap. NASA has an excellent safety record with only a few mishaps.

Hazards

Fires

Hydrogen is invisible, odorless, and not easily detected. To prevent problems, a small amount of colorant and odorant is added to the gas.

For any fire to occur, a proper mixture of hydrogen and oxygen and the minimum ignition are required. Compared with gasoline and methane, hydrogen represents the greatest detonation hazard because of its wide flammability range and low ignition energy. When leaks occur, the lower limits of flammability and detonatability determine whether a fire will occur. One consequence of hydrogen's high burning velocity, compared with gasoline and natural gas, is that hydrogen has a higher probability of undergoing a simple combustion rather than an explosion.

Hydrogen's high diffusion velocity and high buoyant velocity may help prevent fires. In unconfined areas, hydrogen leaks will rise and diffuse rapidly to noncombustible concentrations. A gasoline leak will remain a fire hazard much longer than will a liquid hydrogen leak. If a leak develops in enclosed spaces, hydrogen quickly spreads to all areas of the room, forming a combustible mixture. To minimize this danger, hydrogen should be handled outdoors whenever possible. Handling hydrogen in confined areas requires adequate ventilation.

Ignition sources must be avoided, which is often difficult with hydrogen systems. Experimental uses of hydrogen often require auxiliary electrical equipment. Sometimes static electricity can cause ignition. When hydrogen is exhausted to the atmosphere, ignition control is impossible because of possible electrical storms.

Hydrogen fires are difficult to locate, extinguish, and avoid, because they are invisible. Because they are nearly impossible to detect visually during the day, ultraviolet and infrared technology has been developed to detect hydrogen day fires. However, this is not much of a problem at night, because flames are visible at night. The advantage of a hydrogen fire is that it burns rapidly and does not last long. Hydrogen fires do not generate toxic smoke as do gasoline fires.

Overpressure

Several key elements comprise a detonation hazard, including: (1) fire, (2) a pressure wave resulting from a dramatic increase in pressure, and (3) shrapnel. Of these components, the pressure wave causes the greatest structural damage.

Overpressure that exceeds 120 psi can have devastating effects. A fuel with a wide range of flammability has an increased chance for detonation that can cause extensive structural damage and injury. Hydrogen has wider combustion and detonation ranges than do gasoline and natural gas, and therefore the greatest range of detonation.

Asphyxiation and Frostbite

Hydrogen as a gas can cause asphyxiation. In an enclosed area, if a large leak develops in a hydrogen container, asphyxiation may occur. Symptoms are dizziness, vomiting, unconsciousness, and even death.

Because liquid hydrogen must be kept very cold, frostbite is a concern. Human skin can easily be frozen or torn if it comes in contact with cryogenic surfaces. All pipes containing cryogenic hydrogen must be insulated, and personnel must insulate body parts that could come in contact with these cold surfaces. Where liquid hydrogen is used, the ground should be paved with concrete and not asphalt, which has combustible properties.

Leakage and Condensation

At room temperature, hydrogen may weaken structural equipment by embrittlement. If hydrogen gas escapes, it can be dangerous in an enclosed area. Low temperatures may also be a cause of hydrogen embrittlement.

Liquid hydrogen is sufficiently cold that if air enters the storage vessel or vent system, the air will condense to a solid and exert zero vapor pressure. This forms a combustible mixture. Because the air is so condensed, a considerable amount of accumulation can result. A large fraction of this air can condense in the vent tube of the storage vessel. If the tube becomes plugged, pressure can build up and result in a catastrophe.

Air condensation can also be dangerous if it occurs externally to the liquid hydrogen system. Cold hydrogen vented through uninsulated pipes will condense liquid air on the outside of the pipe. If allowed to drip on a combustible surface like asphalt, an explosive mixture can result. This dripping phenomenon can cause equipment damage and has cracked deck plates on liquid hydrogen transport barges.

Avoiding Hazards

Leaks

Studies have demonstrated that undetected leaks cause most hydrogen accidents. Because hydrogen is such a small molecule, hydrogen can diffuse readily through joints or connections, even those considered airtight. Leaks can be detected by running inert gases, such as helium, through the system. When a leak is suspected, soapy water helps identify the leak. Whenever possible, the number of joints and connections should be reduced, and joints should be welded.

Some plants use flare stacks to vent leaks into the atmosphere. The stacks allow excess hydrogen to be burned without damaging the system. When a large volume of hydrogen must be vented quickly, flare stacks are generally used.

Because leaks will undoubtedly occur, electrostatic sparks and ignition sources must be eliminated. Friction and mechanical sparks can arise during normal operation, and equipment should therefore be kept at a safe distance from the hydrogen source. Smoking and striking matches, of course, should always be prohibited. Buildings should use lightning rods in case of electrical storms. Any combustible material should be removed from the site.

Purging

Air must be purged from a hydrogen system so that a combustible mixture cannot develop within a hydrogen container. After operation of a system, it is advisable in most cases to purge the remaining hydrogen. Inert gases such as helium and nitrogen are often used for purging. The inert gases pressurize the containers, and the resulting mixture is vented. This process is then repeated. About 8 percent of hydrogen accidents are a result of inadequate purging.

Following a purge, testing for residual oxygen must occur before hydrogen is reintroduced. Gas chromatography accurately tests for trace amounts of oxygen. Samples from various areas of the system should be analyzed to determine whether an adequate purging has occurred.

If a large leak exists, venting may not be adequate to eliminate the hydrogen. This happened in the rocket development program, and an entire room was purged by pumping in nitrogen gas until the concentration of oxygen was below 4 percent.

Pressure Relief

When the pressure of a container exceeds its design limits, a pressure relief device is needed to vent the excess gas. Such a device, a rupture disc, consists of a ring covered with a membrane of predetermined thickness, designed to withstand a maximum pressure. When the pressure is exceeded, the membrane bursts and releases the gas. Venting cannot be stopped until all the gas is released. Rupture discs are not used alone, but as back-up devices to spring-loaded valves.

Pressure relief to prevent an unwanted pressure rise is also necessary for liquid hydrogen when the system finally warms to ambient temperatures. The normal pressure relief is set at a predetermined level, backed up by a safety pressure relief valve, which is backed up by a rupture disc.

Venting

Check valves in the vent line will prevent backflow of air, even when venting ceases. Some vent stacks are equipped with helium supplies that purge excess hydrogen from the vent stacks.

Routine venting of liquid hydrogen storage dewars involves small flows directly to the atmosphere. No flaring for the vented hydrogen gas is required. In liquid hydrogen railroad cars, the boil-off losses are vented in a system that mixes air with the exhausted hydrogen in quantities below the combustion limit. Liquid hydrogen is normally stored at higher pressure so that any leakage will cause hydrogen to escape, rather than air to enter the storage dewar. Storage dewars of liquid hydrogen should be independent from the rest of the system. The normal boil-off losses should also be vented separately from the system.

Embrittlement

Because of hydrogen's unique molecular properties, it can diffuse into materials that are impermeable to most fluid gases. Diatomic hydrogen (H_2) is the most stable form of hydrogen. When exposed to enough energy, diatomic hydrogen can dissociate into hydrogen atoms, which can permeate a container surface because of their small size. This phenomenon is called embrittlement, which weakens the metal lattice, and causes it to lose ductility and eventually it can fracture.

When liquid hydrogen is used, cryogenic embrittlement can also occur. The cold temperatures of liquid hydrogen cause the metal to lose ductility or pliability, and the metal eventually fractures.

Certain metals are more susceptible to embrittlement,

including certain types of steel compared with highly alloyed austenite stainless steel. Copper alloys show degradation only at high temperatures. Aluminum exhibits significant surface cracking, whereas nickel alloys exhibit all the degrading mechanisms.

Hydrogen diffusion into the metal lattice is more likely to occur in the gas stage than in a liquid stage. Liquid hydrogen can be safely stored in double-walled dewars made of an inner stainless-steel core and a carbon-steel vacuum jacket.

Preventing Explosions

The geometry of the hydrogen container plays a role in the transition from early heat and sparks to detonation. Decreasing the length of the pipe and increasing its diameter will minimize the detonation stage. Another method of minimizing explosion risk is to use a lean mixture of hydrogen and air, or a mixture toward the lower end of the flammability range. This greatly lowers the burning velocity, and the flame cannot reach detonation.

If a large fire does occur, only trained firefighters should extinguish it. Methods to extinguish hydrogen flames include water (about 80 percent effective), dry chemical extinguishers (about 86 percent effective), production shutdown (about 81 percent effective), and explosion venting (about 91 percent effective).

Proper Instrumentation

Because hydrogen is colorless, tasteless, and odorless, it is very difficult to detect. Liquid hydrogen is a volatile liquid that is seldom seen. Therefore, the operator is completely dependent upon instrumentation for safe, reliable, and efficient operation of the system. Instrumentation should include gauges that detect temperature, pressure, flow, and hydrogen levels.

ACCIDENTS AND SAFETY STUDIES

NASA has been one of the largest users of liquid hydrogen. It uses highway tankers that carry between 3,000 and 16,000 gallons of liquid hydrogen. Throughout NASA's history, it has experienced a few accidents involving hydrogen systems, both on the highway and during off-loading procedures.

One accident involved a liquid hydrogen truck that traveled too fast around a curve. It went off the road and rolled thirty-six feet downhill. The truck came out of the accident in good shape.

Another accident involved a hydrogen tank that unhitched from the cab and veered seventy-five feet down a ditch. The safety discs ruptured as they were designed to do, and the hydrogen evaporated safely into the atmosphere.

A third accident involved a hydrogen delivery truck traveling at 50 mph. It was hit by another truck, and its tank was not damaged, remaining intact.

Testing at Wright Patterson Air Force Base involved several styrofoam-lined aluminum containers. Some containers stored liquid hydrogen and some stored jet fuel. Bullets were fired into the containers. Hydrogen spilled out through the bullet holes and evaporated. The bullet fired into the jet fuel container caused a shock wave and ruptured the container. The bullets ignited both fuels, but neither container exploded. The jet fuel fire persisted much longer than did the hydrogen fire.

Another experiment involved a 6 million volt generator that simulated lightning strikes on the containers. Fires resulted in both fuel containers, but there was no explosion in either tank.

Studies at the University of Miami tested the relative dangers of hydrogen, methane, and propane gas leaks using kitchen models with gas stoves. In each of the experiments, propane presented the greatest danger and hydrogen the least. Hydrogen diffused straight up the vent without forming a combustible cloud. Methane formed a small combustible cloud, and propane formed a large combustible cloud.

Even after years of hydrogen use, experts cannot agree on whether hydrogen is more or less safe than conventional fuels. Some attributes make it safer, and other attributes make if more dangerous. As with fossil fuels, great care and precaution must be taken when working with hydrogen. Safety hazards do not preclude its use as a fuel employed in vehicle transportation.

Chapter Six

PRACTICAL APPLICATIONS FOR HYDROGEN ENERGY

From Power Plants to Lawn Mowers

The potential uses for hydrogen energy are unlimited. Hydrogen energy can replace every fossil fuel energy source, which means we have a solution to dwindling fossil fuel reserves, not to mention global warming. This also means that solar-hydrogen generators can eventually replace coal-operating power plants, whether through wind power, hydro power, photovoltaics, or solar thermal. Hydrogen's great advantage is that it can be produced locally, meaning that people do not have to be dependent on large utility companies or a select group of oil-rich nations for their energy needs. Fuel cell technology allows individual households to get off the power grid, and with solar-hydrogen electrolysis, they could produce their own hydrogen fuel.

Industries would not be at the mercy of utility companies and fluctuating energy prices. They too could produce solar hydrogen power on site, using unlimited water and sunlight. Energy crises such as the 2001 California blackouts could be avoided.

We have discussed the great potential for the automobile industry. As we will discuss in this chapter, the air, rail, and consumer industries will also benefit from hydrogen fuel. Mass ground transportation will benefit greatly. Prototype hydrogen buses have already been built, demonstrating that the technology has arrived. Some of these fleet prototype facilities produce hydrogen on site through water electrolysis. California is beginning to prepare for zero emission vehicles in the year 2003. Car manufacturers will sell commercial hydrogen vehicles to meet these environmentally friendly standards, so a supporting infrastructure must be established.

Mundane applications for hydrogen utilization in flashlights, lawn mowers, scooters, tractors, household appliances, air conditioning, wheelchairs, and heating sources will occur soon. Research is currently being conducted on micro fuel cells that will be a power source for computers and cellular phones. A wise entrepreneur will be getting prepared. The twenty-first century will present exciting answers to what seemed like insurmountable problems. How fast we evolve toward this new economy will depend on the public, politicians, government, and industry. Urgency is mandatory, as we are running out of time.

AIR TRANSPORT

During the 1986 congressional hearings of the Hydrogen Research and Development Act, Dr. Patrick Takahashi, director of the Hawaii Energy Institute, made a strong testimonial regarding the future of hydrogen in aircraft. He said, "Of all the alternatives available, hydrogen can make the greatest contribution to future air travel, in part aided by having the highest energy content per pound of any fuel." The United States government recognizes the potential for hydrogen fuel in aircraft and has sponsored research for its feasibility.

Research shows a number of advantages. The high specific heat of liquid hydrogen allows the fuel itself to be utilized as a heat sink, to cool engine hot spots with air from the engine cycle. Studies showed that the ratios of the heat of combustion for Jet A fuel compared with hydrogen favors liquid hydrogen by a factor of 2.8. In other words, the specific fuel consumption of a hydrogen-fueled engine would be less by a factor of 2.8. Hydrogen will increase the engine life of an airplane and reduce engine maintenance, because hydrogen is clean and free of impurities that either erode or corrode. Hydrogen is injected into the combustion chamber in a gaseous form so that very rapid diffusion and mixing occurs. This promotes complete and rapid combustion to provide a uniform temperature profile, which minimizes thermal stress in combustion. With a hydrogen-fueled aircraft, harmful emissions would be minimal.

Disadvantages include the low density of liquid hydrogen

compared to Jet A fuel, meaning the potential savings in fuel weight represented by reduction in specific fuel consumption are offset by the 3.8 times larger volume required to contain it. Hydrogen-fueled aircraft must have a different fuel tank arrangement, and therefore a higher drag.

Liquid hydrogen-fueled aircraft will have a low wing loading at takeoff, requiring a higher thrust-to-weight ratio. Because of low wing loading, these hydrogen aircraft could take off with engines throttled, resulting in less noise. Hydrogen-fueled aircraft would result in new levels of quietness.

To keep liquid hydrogen at -423 degrees Fahrenheit would require an insulating system for the tanks and fuel system. Before introducing liquid hydrogen, all air must be removed from the tanks, using helium or CH_2 for purging. These gases do not freeze on contact with liquid hydrogen. Any air introduced with liquid hydrogen would freeze solidly to plug up the system.

Supersonic Transport Aircraft (SST)

In 1975, a Mach 2.7 cruise speed liquid hydrogen airplane was designed to carry 234 passengers 4,200 nautical miles and compared with a Jet A fuel version. The passengers sat in the central section of the fuselage directly over the wing in a double-deck arrangement. Liquid fuel was stored in tanks in front and behind the passenger compartment. No provision was made for physical access between the passenger compartment and flight deck. The fuel system added about 80 percent more weight than a Jet A fuel system, including the weight for insulation and vacuum tubing. The plane had a larger fuselage, but smaller wingspan.

The Jet A fuel system plane, when completely fueled, weighed about twice as much at takeoff. This advantage for the hydrogen aircraft minimizes operating costs since the plane needs smaller engines. The operating weight of the hydrogen plane, when empty, was 77 percent less than the Jet A plane. This 32,700-pound reduction would allow fuel savings for liftoff and acceleration.

A hydrogen plane offers economic benefit. The Jet A fueled plane requires 38 percent more energy than the liquid hydrogen-fueled plane. Development and production costs of a liquid hydrogen plane are significantly less than those of a Jet A plane, and a hydrogen plane has advantages in insurance costs and depreciation as well.

Many environmental benefits would accompany hydrogen planes. They are significantly quieter; side noise is four decibels quieter, and flyover noise is six decibels quieter. The design of a hydrogen airplane reduces overpressure caused by a sonic boom.

In 1971, environmental pollution caused the United States to cancel its supersonic program. Pollution in the stratosphere from exhaust products caused great concern. Because of low levels of air circulation, these exhaust products lingered in the stratosphere. Exhaust products of a Jet A-fueled plane include carbon, nitrogen, hydrocarbons, oxides of carbon, sulfur, and water. The hydrogen plane uses 17.1 pounds of hydrogen fuel per nautical mile, generating 153 pounds of water per mile. A Jet A-fueled plane uses 61.4 pounds of hydrocarbon fuel per nautical mile, generating 76.9 pounds of water per mile. Proper design of the hydrogen/air mixture injector and combustor could minimize emission of nitrogen oxide in the hydrogen plane. Use of hydrogen fuel would also alleviate pollution near the ground during takeoff. Jet A engines emit objectionable odors, but liquid hydrogen fueled engines emit no odors.

Nitrogen oxide acts as a catalyst to reduce ozone to oxygen. Ozone is present in the stratosphere, shielding the earth from ultraviolet radiation. Small amounts of nitrogen oxides can drastically reduce this important stratosphere ozone. This is one reason for the cancelled SST program. To minimize nitrogen oxides, combustion temperatures must be kept low, and duration of the combustion process must be as short as possible. Hydrogen airplanes would produce nitrogen oxide levels significantly lower than those of Jet A planes.

Water vapor in the stratosphere concerns some scientists. A hydrogen plane emits twice as much water vapor as a Jet A aircraft. Clouds of water vapor can affect the reflection and absorption of both ultraviolet radiation from the sun and infrared radiation emitted from the Earth. Water vapor in the stratosphere can react with free oxygen atoms (O), reducing their availability to react with molecular oxygen (O_2) to form ozone (O_3). Many experts believe that a fleet of hydrogen supersonic transport planes would have a negligible effect on atmospheric temperature and would not contribute to global warming. Only a small percentage of water vapor would occur over what is continuously present in the stratosphere as a result of natural occurrences. However, other experts hypothesize that it may contribute to global warming.

Water Vapor

Perhaps this is a good time to look at the potential problems of water vapor. Some scientists raise the concern that hydrogen energy systems would produce more water vapor than fossil fuel systems, possibly contributing to global warming. T. Nejat Veziroglu, Ph.D., professor at the University of Miami, does not believe this to be the case. Because only flame combustion of fuels in air or oxygen produce water vapor, Veziroglu states the solar-hydrogen energy system will produce less water vapor than the fossil fuel system.

He claims that global warming, caused by using fossil fuels, also causes an increase in water vapor generation. Since the beginning of the Industrial Revolution, the Earth's mean temperature has increased by 1 degree Fahrenheit. He calculates the additional water vapor generation by global warming was essentially caused by combustion of fossil fuels. Just as water evaporates faster on a hot day, global warming increases water evaporation. His studies have concluded the following:

1. Fossil fuel systems would generate much more water vapor than solar-hydrogen systems.
2. The additional water vapor generated by global warming is much greater than that produced by the combustion of fuels.
3. The amount of water vapor generated by fuel combustion is minimal, compared with that generated naturally.
4. The solar-hydrogen system causes the smallest increase in water vapor generation.

Because it creates less water vapor generation, the solar-hydrogen system is more compatible with the environment than using fossil fuels. We can take this a step farther. If we converted to an entire hydrogen economy, global warming would decrease, temperatures would decline, and therefore water vapor levels would decrease.

Subsonic Aircraft

Researchers have also investigated using hydrogen as a fuel for subsonic aircraft. One design for a liquid hydrogen plane included passenger space for 400 people, with a range of 5,500 nautical miles at a cruising speed of 0.85 Mach (650 mph). Again, liquid hydrogen storage in the fuselage left no access between the passenger compartment and flight deck. Because a high temperature heat shield is not required for a subsonic aircraft, the outer tank covering can consist of simple fiberglass, which provides mechanical protection for the vapor barrier and plastic foam insulation. Both liquid hydrogen and Jet A-fueled subsonic airplanes require the same basic type of turbine engine. The cryogenic hydrogen aircraft provides a large heat sink that effectively cools the turbine.

The economic advantage of using hydrogen fuel in aircraft is that it is a high-energy fuel, resulting in 38 percent lighter gross weight than a Jet A-fueled airplane design. Hydrogen-designed aircraft could operate from shorter runways. A Jet A airplane requires 12 percent more energy to transport 400 passengers 5,500 nautical miles at Mach 0.85 speed than does a hydrogen aircraft. A liquid hydrogen plane would be lighter in net weight than a Jet A aircraft, thus requiring smaller engines. A hydrogen plane would cost slightly more to deliver and produce. Of course, environmental advantages would be much greater with a hydrogen plane.

The advantage of using liquid hydrogen fuel in subsonic transport generally increases with the amount of energy required to perform the mission. For example, after a 130-passenger Mach 0.85 aircraft travels 1,500 miles, the economic advantages of hydrogen would manifest. For a 400-passenger Mach 0.85 aircraft, the economic benefit occurs at 2,000 miles.

RAILROADS

The U.S. Department of Energy studied the feasibility of converting both diesel and electric locomotives to hydrogen operations. They looked at three storage options: (1) iron-titanium metal hydrides, (2) cryogenic liquid hydrogen, and (3) 3,500-psi compressed gas.

Locomotives use a turbocharged diesel engine to generate electric power for traction motors. Engine assemblies come in eight-cylinder configurations, but can be incremented in two-cylinder groups of up to twenty. Power output ranges from 1,000 to 3,000 horsepower.

The study concluded that converting an existing engine to hydrogen, using liquid hydrogen and high pressure injection, would provide certain advantages:

1. The locomotive power levels would increase from 15 to 20 percent due to basic cycle changes with hydrogen. If a series turbo charged air-charge cooling system was used, the engine would gain another 15 to 20 percent in power.
2. The overall thermal efficiency would increase by a maximum of 5 percent if the engine utilized exhaust recovery.
3. A hydrogen train would eliminate smoke, odors, carbon monoxide, and unburned hydrocarbons. Oxides of nitrogen would remain, but newer technology can drastically reduce these emissions.

Development of a hydrogen locomotive would free the railroads from dependence on fossil fuels. Liquid hydrogen storage cars have been in use for many years, and the industry currently uses a 28,000-gallon tank car. Today, liquid hydrogen is significantly more expensive than diesel oil, but as the hydrogen economy develops, the price will become more competitive.

In summary, the study concluded that locomotive diesels can be converted successfully to hydrogen. After conversion, engines will have higher power output with moderate efficiency improvement and longer engine life. The engines would be nonpolluting, except for nitrogen oxide.

DIESEL ENGINES

Nearly 90 percent of the world's trade traffic is accomplished by diesel engines, because diesel engines are more efficient than internal combustion engines. Diesel engines can burn almost all liquid and gaseous energy sources, which can easily be taken along by the transport system themselves. This includes hydrogen. Diesel engines are used in ground transport vehicles, locomotives, and seafaring vessels. Diesel engine vessels comprise more than 90 percent of the intercontinental sea freight transport ships.

Using hydrogen in diesel engines will reduce the amount of carbon dioxide and other greenhouse gases. The direct injection of hydrogen into these supercharged diesel engines produces only marginal emissions of carbon dioxide, carbon monoxide, and unburned hydrocarbons. These are formed by the combustion of small quantities of lubricating oil.

Special tankers are being designed today to provide the infrastructure for hydrogen transport. The European Community will purchase hydrogen produced from hydroelectric-power electrolysis in Quebec, Canada. The Euro-Quebec project is focusing on two methods for transporting liquid hydrogen: (1) the development of dedicated ships and barges, and (2) the development of containers with better insulation techniques. The WE-NET project in Japan is also exploring the development of dedicated tankers for transporting liquid hydrogen.

TWO-WHEELED VEHICLES

Researchers at Energy Conversion Devices, Inc., in Troy, Michigan, studied hydrogen applications for two-wheeled vehicles or motorbikes. Automobiles create the most air pollution in developed countries. However, two- and three-wheelers are responsible for most pollution in the rest of the world. Several factors contribute to this problem. Until recently, most countries had no government emission standards and had poor quality fuel. These developing countries have an enormous number of two-wheelers, most are not regulated, and many are poorly maintained. All of these factors lead to poor air quality.

The global motorized two-wheeled fleet exceeds 200 million vehicles. By 2010, this number is expected to grow to 500 million. Asian markets represent 80 percent of two-wheelers. Most of these large Asian markets have extremely poor air quality, but they also have lots of solar energy. This provides a promising market for hydrogen-powered scooters fueled by photovoltaic electrolytic hydrogen generators. Because of their large affluent class, the cities of Delhi, Shanghai, and Taipei are especially good markets.

Energy Conversion Devices' study projected that 1 million hydrogen-powered internal combustion engines would displace millions of barrels of gasoline over their lifetime. They calculated that 120 million tons of carbon monoxide and 90 million tons of hydrocarbons would also be displaced.

FUEL CELL WHEELCHAIRS

In Florida, a hydrogen fuel cell-powered electric wheelchair was developed to offer cost competitive benefits over current lead/acid battery technology. The project plans to showcase six hydrogen fuel cell wheelchairs.

The benefits include a lighter weight chair, by thirty pounds. Lighter weight means longer battery charge, less wear and tear, and the option of using less expensive lifting equipment. Lead/acid batteries take up to eleven hours to recharge, in contrast to a hydrogen fuel cell that takes only minutes to charge a hydride canister. Batteries produce less power toward the end of the power cycle, whereas a hydrogen fuel cell runs as long as the hydrogen supply lasts.

COOKING FUEL

Cooking fuel is a major energy need for large populations of people in the world with no access to electricity or fossil fuels. Many of these people use scraps of wood, coal, kerosene, and cow dung as cooking fuel. These fuels produce poisonous smoke, causing major health problems and environmental damage.

In India, 80 percent of the urban population and 90 percent of the rural population use some type of organic feedstock for cooking. Few have access to propane or kerosene. In China, coal is the major fuel for more than 80 percent of the population.

Renewable hydrogen produced from photovoltaics (to be discussed in chapter nine), biomass, and other resources can meet the energy needs for most of the people in developing nations. In rural areas, a small integrated hydrogen system (perhaps 1 kilowatt) used mostly for cooking can meet the needs of a single household. Catalytic hydrogen burners have been developed. A very important application will be using solar hydrogen for water pumps, a godsend for most rural communities.

Energy Conversion Devices, Inc. has developed a series of metal hydride hydrogen storage systems, the output ranging from a few watts to several kilowatts. These systems can be used for portable fuel cell applications and for hydrogen internal combustion engines. They can also be used for hydrogen cooking.

Renewable hydrogen energy use in developing nations will help solve global environmental problems. Quality of life would begin to improve for more than 2 billion people not connected to the electric grid.

DOMESTIC USES

Using hydrogen as a residential fuel can be justified, if hydrogen gas can be produced, delivered, and used by residential consumers more economically than natural gas, oil, coal, and electricity. Only when fossil fuel prices become too high or when shortages develop will substitution with hydrogen take place. This scenario will eventually develop, as it is just a matter of time. During the transition period, it will be necessary to convert applications that currently use fossil fuels to hydrogen uses. The distribution network already in place has equipment potentially capable of distributing hydrogen to homes and industry after some modification. Hydrogen distribution could occur through the natural gas distribution network that serves most of the country, as natural gas constitutes more than one-quarter of the total energy supply in the United States.

When fossil fuels become scarce, energy consumers will be faced with the option of converting their appliances from the present energy source to a new one. These consumers will have to choose which energy appliances to buy.

Space heating accounts for 65 percent of the total energy consumed in a household and is the largest energy-consuming application. The quantity of energy used to heat water usually depends on the number of occupants in a house. Energy for hot water heaters equals about 13 percent of total energy consumption. Cooking consumes 3 percent of the total energy, whereas clothes-drying accounts for 17 percent of the total annual energy consumption.

If hydrogen is to replace natural gas in residential appliances, the behavior of hydrogen in burners designed to combust natural gas must be examined. Modifying the burner is necessary, as hydrogen cannot be substituted directly into residential appliances. Hydrogen has a number of properties that could be exploited in catalytic burners, rather than flame burners.

The burners built into domestic gas appliances are designed to burn fuel from a low-pressure gas source, called "atmospheric burners."Ambient air enters the tube through adjustable openings and is transported by the high-velocity gas stream.

For a burner operating on hydrogen to deliver the same amount of heat per unit time as one operating on natural gas requires 3.25 times as much fuel. Without changes, the current burners will allow passage of 2.9 times as much hydrogen. The hydrogen burner will deliver about 10 percent less heat per unit time. The air/fuel ratio for complete combustion of natural gas is approximately 10:1, compared with a ratio of 2.4:1 for hydrogen.

The burning velocity of hydrogen is 9.2 feet per second, compared with 1 foot per second for natural gas, causing potential flashback problems in hydrogen burners. To prevent flashback, the gas velocity must be nine times greater than for methane. The problem of flashback in converting natural gas appliances can be addressed in two ways: (1) simultaneously increasing the upstream pressure and decreasing the port size, or (2) decreasing the amount of primary air so that the mixture upstream of the burning ports is not flammable.

Currently, the American National Standards Institute has only one criterion for combustion pollutants from domestic gas appliances, and that regards carbon monoxide. Hydrogen uses no carbon, and therefore produces no carbon monoxide. However, the open flame of a hydrogen burner produces significantly larger quantities of nitrogen oxide than does natural gas.

The easiest way to convert existing natural gas appliances to hydrogen would be to design a new burner head. To ignite the proposed hydrogen burner, catalytic ignition is more practical in terms of fuel expenditure and air pollutant control. When the initial flow of hydrogen passes over a catalyst of noble metals, hydrogen reacts with oxygen in the air at room temperature and thereby produces heat. When the surface temperature reaches the auto-ignition temperature of the hydrogen/air mixture, a flame is initiated. The ignition noise will not be completely eliminated due to hydrogen concentrations in the vicinity of the burning ports and ignition.

In summary, the most serious drawbacks of unmodified burners operated with hydrogen rather than natural gas will be flashback, noise, and nitrogen oxide emission. Most of these problems can be remedied by replacing these burners with hydrogen catalytic burners.

UTILITIES

Utility companies will play a major role in the transition to a hydrogen economy. The required new storage and transportation systems will have to meet the needs of the customer. Because hydrogen has stationary as well as mobile applications, the energy carrier hydrogen offers great flexibility in our evolving energy system.

Stationary applications of hydrogen will remain within the present domain of utility companies. Because hydrogen is an energy carrier, its mobile application can also be handled by utility companies as they enter into the fuel business.

The production of hydrogen by electrolysis offers many benefits to power companies. Electrolysis of water to produce hydrogen can be used for frequency control or load management. This allows power companies to save this energy as reserves in the form of hydrogen. They will be able to tap into these reserves at peak times, allowing for better efficiency and higher revenues. During low-load periods, hydrogen can be produced and reconverted to power at peak loads.

Fuel cells seem to be the most promising technology for the future of hydrogen and can be supplied with either reformed natural gas or directly with hydrogen. The fuel cell represents an ideal bridge between the fossil fuel system and the future electricity/hydrogen economy. Power utilities can enter into the fuel business by generating hydrogen through electrolysis and supplying fuel cells with it. Because fuel cells are used in mobile applications, utility companies can deliver hydrogen to distribution facilities. Fuel cell-powered automobiles will be common in the future, and utility companies will need to supply hydrogen to refueling stations. Home-use fuel cells, currently distributed by utility companies, will need to be supplied with hydrogen fuel.

In remote areas with a weak grid system, utility companies could establish a local hydrogen production system. Through water electrolysis and a subsequent hydrogen storage system, a cogeneration power plant could provide hydrogen to fuel cells or internal combustion engines to supply the customer with power.

Earlier, we discussed the various fuel cells. High-temperature fuel cells are suitable for large-scale stationary application, producing power in the range of 1 megawatt to 20 megawatts. Still in development, market penetration will not occur before 2005. For small-scale stationary applications, low-temperature PEM fuel cells are appropriate. This technology is supported by billion-dollar research programs at DaimlerChrysler, Ford, General Motors, and Toyota.

Two German utility companies, HEW and HGW, operate two phosphoric acid fuel cell (PAFC) operations in the urban area surrounding Hamburg. The power of each fuel cell is 200 kilowatts. Besides the PAFC, PEM fuel cell technology is sufficiently developed for small prototype power plants. In 2000, Ballard provided a 250-kilowatt fuel cell unit to a consortium of European utilities for a natural gas cogenerating power plant in Berlin.

Almost every application of fossil fuel energy can be substituted with hydrogen energy. Fossil fuels will remain with us during the transition, as they are a source of hydrogen. During the transition period, utility companies will wean themselves from fossil fuels to renewable sources with the production of hydrogen from electrolysis.

Chapter Seven

HYDROGEN'S CURRENT STATUS

The Technology is Available

As we have seen, the government, a few politicians, and industry now realize the great potential for hydrogen energy. However, the general public remains uninformed, and without their support, hydrogen energy will remain in the background. During the 1990s, the public did not worry about energy needs. Fuel prices were low, fuel was plentiful, and the economy had never been better. With the new millennium, concerns about higher prices for heating fuels began to spread across the nation. California began to experience blackouts. Gasoline prices skyrocketed. The public's attention to energy prices is once again bringing alternative fuels to the forefront. Many auto manufacturers have announced the future introduction of hydrogen cars, planting the seed of things to come. Regulations in the state of California have stimulated industry's interest in hydrogen technology.

CLEAN AIR REGULATIONS IN CALIFORNIA

California has set higher standards than any place in the world in their determination to eliminate air pollution. Presently, California has the worst air pollution in the United States and has been the proverbial "canary" for U.S. pollution. Realizing air pollution's effect on the health and well-being of its people, California set in motion some of the most stringent regulations regarding vehicle emissions. To meet these hallmark standards, the auto manufacturers have had to be innovative to maintain their market share. At first, electrical vehicles seemed to be the answer for meeting zero-emission standards established by the state. However, problems such as limited driving range associated with electric vehicles were not consumer friendly, and the auto manufacturers looked at hydrogen vehicles. This is the California story.

The California Air Resource Board (CARB) is responsible for current standards for automobile emissions and gasoline. They have established the cleanest standards in the world. Today, hydrocarbon emissions from a 1999 California light-duty vehicle are on average 50 percent lower than in the rest of the United States. A new California motor vehicle is more than 95 percent cleaner than a new vehicle was in the 1970s. No major power plants in California burn coal or petroleum products, and most power plants in the state use natural gas. During the last twenty-five years, CARB regulations have pushed vehicle manufacturers to pioneer technology resulting in catalytic converters, computerized fuel controls, fuel injection systems, and onboard diagnostics to reduce emissions in conventional motor vehicles. The board passed a regulation requiring the marketing of zero and near-zero-emission vehicles beginning in 2003.

Until November,1998 the regulation stated that beginning in 2003, 10 percent of light-duty vehicles marketed by manufacturers must be zero-emission vehicles. The regulation was amended and now states that manufacturers will have to meet 40 percent of their 2003 requirements of zero-emission vehicles. For the other 60 percent of the emission vehicle requirements, manufacturers could receive partial zero-emission vehicle credit by marketing vehicles with fuel cells using onboard fuel reformers, hybrid engines, and other near-zero emission technologies. The manufacturers would have to provide a 150,000-mile warranty for the emission control system. These CARB zero-emission regulations apply to manufacturers who market 3,000 or more vehicles per year in California.

When the regulation was originally passed in the 1990s, CARB had electric cars in mind. However, fuel cell technology developed rapidly and CARB wanted to encourage manufacturers to explore this technology to the fullest. The state is working with automobile manufacturers to demonstrate the viability of fuel cell vehicles, as they realized that fuel cell vehicles would help vehicle manufacturers to meet these stringent requirements. Fuel cell vehicles would be more attractive to consumers because they offer clean operation with the convenience of quick refueling, extended driving ranges, and they operate more efficiently and cleanly on hydrogen.

The population of California in 2010 is projected to be 41 million people, and vehicle miles traveled in the state are projected to increase by 24 percent. CARB believes hydrogen technology, mainly in the form of fuel cells, holds great promise for widespread application in both mobile and stationary sources.

CARB realizes the commercialization of vehicles attaining zero emission or near-zero emissions will be necessary before California can provide healthful air for its citizens. For most of the 1990s, conventional wisdom held that battery-powered vehicles were the only zero-emission vehicles that would be ready for commercialization.

Six major manufacturers have marketed battery-powered electric vehicles in California since 1996. General Motors' EV-1 was the first, unveiled to the public in December, 1995. During the past several years, great advances have occurred in fuel cell and hybrid electric technologies. Some of these vehicles were not truly zero-emission vehicles. CARB realized they needed to recognize this new technology, and in 1998 they amended the regulation for near-zero emission vehicle technology.

Under CARB's current regulation, between 40,000 and 100,000 pure zero-emission vehicles will be marketed in California, beginning in 2003, plus another 300,000 near-zero emission vehicles. After reviewing the regulations in 2000, CARB will have the power to amend them. CARB wants to ensure that the type and quantity of vehicles produced are consistent with projected consumer demand for vehicles that use this technology.

Ford, DaimlerChrysler, Honda, Toyota, and Nissan have all been marketing battery-powered vehicles in California. Most of the test vehicles use a nickel-metal hydride battery offering a range of ninety-six miles. Nissan has a lithium-ion battery with a range of forty-eight to sixty-six miles. Both cost and limited range pose significant problems.

In 2000, Toyota marketed a hybrid electric vehicle (Prius) in the United States similar to the Prius introduced in Japan in 1999. It provides the most efficient use of gasoline and electric engines. Honda has also developed a gaseous-electric car, called the Insight. Several U.S. hybrid cars use diesel fuel.

Nissan's Sentra CA is the first gasoline-fueled car to receive CARB's rating of superultra low-emissions vehicle certification and partial zero-emission vehicle rating. Honda has produced a similar vehicle.

CARB states that the potential of hydrogen-fueled cars is one of the most exciting developments in the air quality field in recent years, especially with the rapid development of fuel cells in light-duty motor vehicles. Even as late as the mid-1990s, fuel cells were viewed as too bulky for use in light-duty vehicles. Recent technology permits the smaller and lighter fuel cell design that can effectively power a car. Alan Lloyd, a member of CARB, said the proton exchange membrane (PEM) fuel cell is ideal for light-duty vehicles because it operates at relatively low temperatures (80 degrees centigrade) and quickly adjusts to various demands for power. He states that fuel cell vehicles can offer the best features of both electric and conventional motor vehicles. Fuel cell vehicles use an electric motor to provide nearly the same emissions as a battery-powered electric motor. By using available fuels, fuel cells could offer the same convenience as does the gasoline vehicle. The vehicles have the potential to produce little or no pollution with hydrogen fuel. Fuel cell vehicles using onboard hydrogen produce no polluting emissions and are pure zero-emission vehicles. Other fuel cell vehicles carry reformers that derive hydrogen from methanol, ethanol, gasoline, and natural gas. These fuels can be distributed within the existing fuel infrastructure at reasonable expense.

CARB realizes that onboard hydrogen would be ideal, but the hydrogen infrastructure is not available to the public. CARB is participating in a demonstration project involving a fleet of fuel cell vehicles using onboard hydrogen. Fuel cell buses in California would greatly benefit air quality, because heavy-duty diesel bus engines are a major source of nitrogen oxides, fine particles, and air toxins.

In 1999, CARB joined with other private and public organizations that had an interest in fuel cells to form the California Fuel Cell Partnership. This partnership will demonstrate and test performance of fuel cells in about seventy-five vehicles. They will also operate a fueling infrastructure. The partnership includes DaimlerChrysler, Ford Motor Company, Honda, Nissan, Volkswagen, BP Amoco, Shell Hydrogen, Texaco, Ballard Power Systems, California Energy Commission, and the U.S. Department of Energy.

Since the year 2000, the partnership has demonstrated ten passenger cars and up to five buses. When the project concludes in 2003, an additional forty passenger cars and twenty buses will have joined the demonstration. DaimlerChrysler's NECAR 4 and Ford's P200 Prodigy will be two of the passenger vehicles used in the demonstration. Current plans are for the demonstration fleet to use onboard hydrogen. The demonstration project will gather important data on the maintenance needs and costs for fuel cell vehicles. CARB believes that maintenance costs of fuel cell vehicles will be lower than for internal combustion engines because they have fewer moving parts. The major concession is the significantly higher cost of fuel cell vehicles.

CARB also recognizes that fuel cell technology can provide electricity to homes, offices, and other facilities. The California Air Resource Board has accomplished much with its high standards, standards that will greatly benefit the world. We tip our hats to the efforts of CARB.

CLEAN AIR NOW

Clean Air Now (CAN) is developing a "hydrogen corridor" in southern California, helping to pave the way for the California Fuel Cell Partnership (CFCP), whose sole purpose is to put fuel cell-powered vehicles on the road. A series of projects will develop a hydrogen fuel infrastructure that supports hydrogen-powered vehicles along a 150-mile stretch from the Pacific Ocean to the desert east of Los Angeles. CAN is responsible for the solar hydrogen-generating facility and Hydrogen Powered Vehicle Project at Xerox Corporation.

Approximately 70 percent of toxic air pollution in the region is caused by diesel exhaust emissions. These emissions contain certain carcinogens that have resulted in increased death rates for citizens living in this area. In response to these concerns, the SunLine Transit Agency hydrogen project has been implemented. The project is researching hydrogen buses, golf carts, and U.S. Army experimental hydrogen trucks.

The project involves producing hydrogen on-site through electrolysis, with storage in nearby pressurized gas cylinders. Fleet vehicles have onboard carbon fiber storage tanks. CAN and SunLine are the first organizations in the United States to demonstrate a 100 percent pollution-free transportation system. CAN believes that replacing heavy-duty vehicles with nonpolluting fuel cell vehicles will eliminate the cancer epidemic caused by diesel fuel.

The hydrogen fuel generated on-site is produced from sunlight and water through electrolysis. Both the generation and compression systems were 100 percent off-the-grid systems. This photovoltaic electrolytic system had an electrical average power value of 2,500 amps at fifteen volts that produced 320 standard cubic feet of hydrogen per hour. The high-pressure storage systems could store 15,000 cubic feet of hydrogen at 5,000 psi.

CAN is also involved with the hydrogen project at the University of California at Davis, which will research two hydrogen buses and three hydrogen internal combustion engine vehicles.

CHICAGO TRANSIT AUTHORITY

In the spring of 2000, Ballard Power Systems, XCELLIS Fuel Cell Engine, Inc., and the Chicago Transit Authority (CTA) completed a two-year demonstration of three Ballard fuel cell-powered buses. The buses logged more than 5,000 hours and carried more than 100,000 paying passengers. Data gathered were integrated into improveed designs for future XCELLIS fuel cells. XCELLIS Fuel Cell Engine, Inc. is a joint venture of DaimlerChrysler, Ford Motor Company, and Ballard Power Systems.

Air Products and Chemicals Incorporated (APCI) provided gaseous hydrogen delivered by cryogenic trailer. It was vaporized for use at a refueling center constructed by APCI at the Chicago Transit Authority Pulaski hub. A fifteen-minute fill-up time provided enough hydrogen for a full day's run of about 250 miles.

THE BMW PROJECT

Since 1979, the German car manufacturer BMW has been one of the leaders in hydrogen research. BMW engineers saw more potential in internal combustion engines fueled by liquid hydrogen than in fuel cells. BMW partnered with the Bavarian State Government and others in 1986 to establish the Bavarian Solar Hydrogen Company (BSH). The company examined the necessary steps to manufacture hydrogen for application in pilot facilities.

The production plant installed 8,300 solar panels over an area of 4,200 square meters at the BSH site in the Upper Palatinate region of Germany. These solar panels could generate 390 megawatt hours of electricity per year, equivalent to supplying electricity to 130 households. From this electricity, the company electrolyzed water to produce hydrogen and oxygen. The oxygen and hydrogen gases were then compressed, and the hydrogen was used in heating facilities, fuel cells, and transport systems. A liquid hydrogen filling station was built on-site, allowing a car to refuel in less than three minutes.

After twelve years, the project concluded, with many of the hydrogen systems still available in prototype form. However, to produce these prototypes competitively in the commercial market remained a long way off. To break into the marketplace, the company concluded that both investment and operating costs would have to be reduced significantly. More than 124,000 people visited the site to witness the beginnings of a solar hydrogen economy.

The Bavarian government was determined to convert the whole apron traffic system at the Munich Airport into a hydrogen-fueled facility. The Munich Airport project uses a 7 Series BMW car that runs on liquid hydrogen to transfer people on the apron. It also uses three low-floor hydrogen buses to transport airline passengers. As part of the project, the world's first public liquid hydrogen filling station was opened at the Munich

Airport in the spring of 1999. A fully automated robot fills the vehicle tank with liquid hydrogen. Through its cooperation on the project, BMW has acquired valuable knowledge regarding robotic service station attendants. BMW's hydrogen vehicles are fully compatible with the tank robots. A driver inserts a chip card into a terminal slot to start the filling process. At this time, we do not know whether robots can be programmed to clean windshields.

In 2000, the Munich Airport filling station sold a liter of liquid hydrogen for 1.2 DM ($0.52) or just over two dollars a gallon. Because of hydrogen's low volumetric energy, hydrogen vehicles can consume 3.5 times more fuel by volume than gasoline-fueled cars, costing ten times the price of gasoline.

BMW has manufactured a fleet of 7 Series cars with twelve-cylinder engines fueled with hydrogen. These cars have lost no refinement in driving dynamics as compared with BMW's other powerful internal combustion engines. During combustion in the spark-ignition engine, the hydrogen oxidizes in the air and becomes water.

BMW has shunned fuel cells to power their vehicles. After ten years of research, they did not think a fuel cell vehicle could compete with the internal combustion engine. Engineers found too many problems relating to power density, water and heat management, platinum requirements, and system costs. Since their research, many remarkable advances have occurred with fuel cells in the last few years. However, BMW has found it can use fuel cells in other ways. It has used a fuel cell to keep a vehicle at a sustainable temperature without running the engine. By using a 5-kilowatt fuel cell generator, it can power an air-conditioning system when the vehicle is stationary. This small unit can supply the entire onboard electrical requirements of the car more effectively than can the present system using lead/acid batteries. This concept unit is called an "auxiliary power unit."

The BMW cars have a 5.4-liter V12 engine with sequential fuel injection. They can attain speeds of over 130 mph. Fuel consumption is 2.5 kilometers per liter, or 5.5 miles per gallon. These cars have a 140 liter-fuel tank (about 39 gallons). At the rear of the car, a fuel cell is mounted for a permanent onboard electricity supply. Gasoline tanks are also installed on the car. The driving range is about 210 miles with hydrogen and about 360 miles with gasoline.

In BMW's hydrogen car, the throttles remain open almost continuously, but the engines run unthrottled with excess air, and are controlled by the amount of fuel supplied to them. The liquid hydrogen achieves a power density 10 percent higher than that of gasoline.

BMW does not think the cars will be more complex than today's vehicles. Liquid hydrogen tanks are certainly more expensive because of the double-walled design and enclosed piping system needed. The three-way catalytic converter will be eliminated, but a simple catalytic converter will be needed for nitrogen oxides. BMW crash tests have demonstrated that hydrogen cars are as safe as gasoline cars.

HYDROGEN IN JAPAN

The Shikoku Electric Power Company in Japan has developed the world's first phosphoric acid fuel cell that uses hydrogen produced with a saltwater electrolysis process. By May 2000, the fuel cell had accumulated 24,358 hours of operating time. The company has explored the possibility of building hydrogen energy systems for remote, isolated islands.

Because of Japan's high interest in hydrogen, it developed the hydrogen energy system technology project referred to as WE-NET (World Energy Network). The project's goal is to build a clean global energy network using renewable energy sources from areas blessed with hydropower, sunlight, and wind power to produce hydrogen and supply it to Japan. Between 1999 and 2003, WE-NET's second phase of research and development will emphasize the technology for hydrogen dispersion. Hopefully, this will accelerate the introduction of a hydrogen infrastructure where hydrogen has the potential to show its superiority.

WE-NET plans to develop two types of hydrogen refueling stations: natural gas reforming and water electrolysis. The water electrolysis hydrogen station has the potential for generating a new night-only service load for the electric utilities. Electric utilities have long awaited such a use for their nocturnal electricity load. Presently, the production cost of water electrolysis of hydrogen is about 1.7 times higher than natural gas reforming for hydrogen production. If the water electrolysis supply increases the night-only service load and promotes load leveling, the electricity costs will be reduced, and they will lower the hydrogen cost. Japanese electric power companies welcome the early arrival of the hydrogen economy as a tool for load leveling. Electric utilities have had to cope with steady growth in demand for electricity, the gap between daytime and nighttime demands, and gaps for seasonal demands. The low range for load in Japan is between 54 and 60 percent, compared with about 60 percent in France, Germany, and the United States.

Japan is making plans for a hydrogen economy and has projected the number of hydrogen vehicles during the next thirty years, along with a projected number of fueling stations.

Year	Number of Vehicles	Number of Stations
2000	640	20
2010	35,500	500
2020	1.84 million	3,000
2030	8.68 million	6,000

According to projections, by 2010, hydrogen freight vehicles will be introduced. At that time, additional hydrogen production plants will also be needed.

ICELAND

Iceland has no fossil fuel energy reserves except for a few peat reserves. However, the country possesses abundant amounts of hydroenergy and geothermal energy, which supply 58 percent of Iceland's energy needs. Only about 15 percent of the hydroelectric potential has been harnessed and about 1 percent of its geothermal energy. Approximately 90 percent of Iceland's buildings are currently heated by geothermal water.

Iceland has a large fishing industry that uses fossil fuels,and that, combined with transportation needs, accounts for 28 percent of Iceland's total energy consumption and 70 percent of its total liquid fossil fuel imports.

In February, 2000, the Icelandic Hydrogen and Fuel Cell Company, Ltd. was founded. The company sought to investigate the potential for eventually replacing fossil fuels in Iceland with hydrogen. As a result, Iceland would become a pilot country for demonstrating a hydrogen economy. This company is partly owned by DaimlerChrysler, Shell International, and Norsk Hydro.

Plans are to operate PEM fuel cell buses developed by DaimlerChrysler in the capital city of Reykjavik, and hydrogen cars by DaimlerChrysler will be introduced into all Iceland. The buses (called NEBUS) and the cars (called NECAR 3) will be mass produced in the year 2004. Pressurized hydrogen gas will be stored onboard the NEBUS in sufficient amounts to operate the bus for 150 miles, the same distance Reykjavik buses presently run each day. Because a hydrogen infrastructure is not in place, the PEM fuel cell for private cars will run on methanol in the transition stage. PEM fuel cells will also power fishing vessels.

Replacing imported fuels with methanol would reduce anthropogenic carbon dioxide emissions by 55 percent. As the hydrogen infrastructure develops, hydrogen PEM fuel cells will replace methanol PEM fuel cells. Iceland's transformation into a full hydrogen economy will be complete by the 2030s.

Iceland has chosen to be the world's harbinger of a hydrogen economy. Both Japan and Germany are making plans for this transformation. California is leading the United States toward hydrogen energy, and by the year 2003, many hydrogen cars will travel the freeways of California. Knowledgeable scientists believe that hydrogen will meet the world's future energy needs. As with any new technology, a transition period will occur.

Chapter Eight

MAKING THE TRANSITION TO HYDROGEN

Government, Economics, and Timetable

The transition to a hydrogen economy has already begun. Rapid advances in technology during the past several years have made a future hydrogen economy highly probable. Powers in government and industry have acknowledged the environmental dilemma and the future fossil fuel shortage. Without the support of auto manufacturers, oil companies, utility companies, the public, and government, it will be impossible to make this transition. Not all these players are enlightened to the advantages of hydrogen, but as the transition progresses, they will come aboard.

How quickly the transition takes place will depend on government policy and public support. A vast majority in both sectors knows very little about the hydrogen economy. When the California zero-emissions policy goes into effect in 2003, public knowledge about hydrogen will escalate. During the transition period until 2003, industry will make major decisions about developing the future hydrogen infrastructure.

GOVERNMENT

The U.S. Department of Energy (DOE) has come to the conclusion that hydrogen will be our future energy carrier. During Senate hearings on the Hydrogen Future and Hydrogen Fuel Cell Commercialization Act of 1995, Allan Hoffman from the DOE testified. He stated,

> "The Department of Energy recognizes the potential offered by hydrogen and embraces the vision that hydrogen will join with electricity in the twenty-first

> century as a primary energy carrier in the nation's energy future. The DOE believes hydrogen will ultimately be derived from renewable resources and be delivered to end users for a variety of applications. Hydrogen will offer a nonpollutant, potentially efficient, and cost-effective energy system that could be derived entirely from domestic energy sources."

The DOE has financed hydrogen research, serving as a liaison between science and industry. Ford Motor Company, DaimlerChrysler, and General Motors have all benefitted directly from its support, and hydrogen vehicles will be available to consumers shortly.

State governments, such as California, have established air-quality regulations more stringent than those of the U.S. Environmental Protection Agency. These zero-emission regulations have forced the major auto manufacturers to develop the hydrogen vehicle. The state of Hawaii has been a leader for hydrogen research through the University of Hawaii. Senator Daniel K. Akaka of Hawaii testified at the 1995 Senate hearings saying,

> "Hydrogen will join with electricity in the twenty-first century as a primary energy carrier for the Nation's energy future. There is growing evidence that hydrogen can be a solution for America's long term energy needs. . . . Now is the time to increase research and development of a new source of energy to minimize the potential of energy supply disruptions and make a smooth transition to the next generation of energy sources. . . . In order to retool for a future fueled by hydrogen, our nation needs active and systemic research, development, and demonstration programs to create the necessary breakthrough so hydrogen can become a viable source of energy."

Sometimes called "Mr. Hydrogen," Senator Tom Harkin has offered wisdom to make this transition to hydrogen. Harkin believes private industry has no financial incentive to take over development of hydrogen energy all by themselves. He suggests that government funding for hydrogen energy will help prepare viable technology to meet the coming oil shortage.

Harkin has recommended "sustainable energy centers." Local center locations would determine the type of sustainable energy used to produce hydrogen. For example, Iowa might use wind and biomass, whereas California could use solar and geothermal sources. The sustainable energy centers would serve as larger scale integration testing facilities. They could observe how hydrogen production, storage, and distribution all fit together. At these centers, the public would be educated and could give their feedback. Hydrogen information would be exchanged in center meetings, conventions, and publications. Harkin said hydrogen technology would create thousands of new jobs, and exciting new business ventures would rise. To speed this transition to hydrogen, the government needs to fund research, educate the public, and provide tax incentives to private industry for research and development of hydrogen technology.

In Europe, countries have formed a Hydrogen Joint Working Group (HJWG) whose goal concerns the hydrogen economic interests of their respective countries. Its purpose is multinational cooperation and exchange of information that will benefit its members. The HJWG plans to assist other countries in developing national associations regarding energy needs.

The time has come to begin the transformation to hydrogen. Hydrogen will allow nations to meet the 1997 Kyoto Protocol formulated at the World Conference on Climate Change. More than 160 nations agreed on steps to reduce emissions of the six most significant gases contributing to the greenhouse effect. The accord allowed flexibility to enable companies and nations to decide when a reduction of emissions is most worthwhile from an economic point of view. In November, 2000 a follow-up United Nations global warming conference was held in The Hague, Netherlands. At the conference, European countries accused the United States of evading its commitment to curb greenhouse gases. Kevin Gurney, a Colorado State University scientist, accused the United States of cheating on the Kyoto pact, saying that the United States wants to use existing forests and farmlands to reduce greenhouse emissions. America is the greatest producer of greenhouse gases, and its proposal in the long run would increase its emissions. In March, 2001, President George W. Bush announced that the United States would not honor the 1997 Kyoto accord of reducing carbon dioxide emissions. Perhaps a hydrogen economy is the only way the United States could meet the agreed upon protocol.

By 2012, the U.S. greenhouse emissions are projected to grow more than 20 percent. Reaching 7 percent below the 1990 levels as the Kyoto agreement states, would require a 30 percent decrease in emissions. Taking matters into its own hands, Portland, Oregon in 1993 became the first city to implement its own carbon dioxide reduction plan.

Setting an example for the rest of the world has been China, who has 11 percent of the world's carbon dioxide output, second to the United States. They cracked down on emissions and reduced greenhouse emissions by 17 percent between 1997 and 1999, demonstrating that carbon dioxide reduction is possible.

TRANSITION FROM FOSSIL FUELS

The most environmentally compatible source of hydrogen comes from electrolysis of water, with solar power as its source. This process is too expensive today, but will be the energy source of the future. The infrastructure for solar hydrogen is not yet in place.

During the transition, hydrogen production will depend largely on fossil fuels, such as Texaco's Cool Water coal gasification plants. They offer an environmentally friendly technique for processing coal, to obtain hydrogen through the gasification process of natural gas and oil. Hydrogen can also be obtained through traditional steam reformation techniques using fossil fuels, the point being, fossil fuels will provide an important source of hydrogen during the transition period. Economic advantages exist as well: the infrastructure is in place; fossil fuel employees will maintain their jobs; investors will continue to profit; and oil companies will become energy companies that build the hydrogen infrastructure as fossil fuels become depleted. When the transition to a full hydrogen economy is complete, enough fossil fuel resources will remain to ensure feedstock for consumer products such as plastics, computers, synthetic fabrics, and so on. It will truly be a win-win situation.

In the early stages of the transition, today's internal combustion engines can be converted to hydrogen. Roy McAlister, founder of the American Hydrogen Association, has developed a kit that converts an average car to operate on compressed hydrogen, methane, or natural gas. The conversion costs between $2,000 and $6,000, depending on the type of car, engine type, and electrical options. Driving range depends on the size of the storage tank. Drivers can easily switch between gasoline and alternative fuels.

As California zero-emission requirements go into effect, hydrogen vehicles will begin to invade the California freeways, requiring an infrastructure to refuel these vehicles. Fuel cells will probably supply power for most of these vehicles, which will operate on either natural gas or hydrogen. Other fuel cell vehicles will carry reformers, from which they can derive their hydrogen from methanol, ethanol, gasoline, and natural gas. These fuels can be distributed within the existing fossil fuel infrastructure at competitive costs, with the advantage of producing near-zero emissions. Many of these transition vehicles will be hybrids, operating on fuel cell technology that will increase the range of electric cars by recharging their batteries.

For liquid hydrogen cars, BMW proposes a three-stage transition to hydrogen energy. During the first stage, compressed natural gas will be used as a fuel. The second stage will use liquefied natural gas, followed by liquefied hydrogen in the final stage. Compressed natural gas vehicles are already state-of-the-art.

As the number of hydrogen vehicles increases, the demand for refueling stations will also increase. Again, we can use the present infrastructure of service stations. The Japanese are convinced that metal hydride containers will supply hydrogen storage for fuel cells. Japan is developing prototype WE-NET stations to refuel hydrogen metal hydride vehicles.

As the transition progresses, hydrogen will be produced from the natural gas source at the processing plant. Pipelines from the natural gas infrastructure will deliver hydrogen to consumers, including service stations. At the service station, the hydrogen gas will be compressed into storage cylinders to refuel hydrogen cars.

Another plan involves supplying hydrogen from central natural gas reformers for distribution by truck or pipeline. To support a future large-volume demand, central natural gas reformers can be built to scale to supply sustainable generation of hydrogen. Currently, hydrogen can be produced and distributed at comparable (and well below taxed gasoline) costs.

Five European firms in Germany, Switzerland, and Austria are researching how to supply hydrogen to consumers. Called the European Hydrogen Filling Station (EUHYFIS), they have determined that a grid connection system that guarantees safe handling in any operating condition will be required for this refueling station infrastructure. The grid will also supply backup electricity when the renewable source cannot produce enough hydrogen fuel. Hydrogen will be produced by electrolysis at the filling stations, the rate determined by the delivery ratio of the compressor. Purity of hydrogen could be as high as 99.999 percent.

For fully commercial application, fast refueling stations will be of utmost importance. Hydrogen will be stored under high pressure, and vehicles will be refueled within a few minutes. EUHYFIS believes that industry and government must work together to prepare for the changeover from fossil fuels, with hydrogen being the logical last step in the energy transition from high carbon to noncarbon energy sources.

The European Hydrogen Network (HYNET) is planning for several alternatives in the transition. One option uses a small hydrogen generator fueling station, and another option generates hydrogen onboard the vehicle.

A technological breakthrough is necessary for the development of a compact and efficient storage system that will revolutionize the hydrogen energy system. Car manufacturers agree that a general purpose hydrogen vehicle that provides an acceptable autonomy to the consumer must store at least 11 pounds (5 kilograms) of hydrogen. Storage of sufficient amounts of hydrogen fuel remains the most challenging task in establishing the hydrogen economy.

ECONOMICS

To completely convert to a hydrogen economy and replace fossil fuels will take high incentive. Some believe it will take the resolve of the Manhattan Project or Apollo program to accomplish this goal. Government, industry, and scientists are all coming to the conclusion that we do not have a choice. Of all the alternative fuels, hydrogen seems to address the most concerns. It offers the great advantage of incorporating existing oil companies and auto manufactures with a whole new industry of solar hydrogen technology. Does Saudia Arabia know something that we do not? It has announced plans to supply 25 percent of the world's energy as hydrogen. The hydrogen story suggests the wise investor might look at the future of hydrogen rather than the "dot-coms."

T. Nejat Veziroglu, Ph.D., from the University of Miami and nominee for the 2000 Nobel Prize for Economics, believes that when one takes into account the environmental damage of fossil fuels, hydrogen is the least costly. He lists great advantages for hydrogen: Hydrogen would (1) reduce carbon dioxide emissions and acid rain, (2) eliminate dependence on oil imports, (3) reduce the trade deficit, (4) provide new jobs for establishing and servicing the new energy systems, (5) make the United States a leader in future energy technology, and (6) assure low-cost fossil fuel supplies for consumer applications.

In the 1985 U.S. Senate hearings, Veziroglu stated that annual environmental damages cost 0.33 trillion dollars, as determined by a three-year University of Miami study. Sixteen years later, this figure is much higher. This sum includes medical expenses related to pollution, loss of working efficiency because of environmental problems, and related environmental disease. The figure also includes the harmful effects of acid rain to lakes and water life, and the harmful effects of ozone on farm produce and plant life. The cost of fossil fuel-related deaths was estimated at $100,000 per person. A recent scientific study reported in the December, 2000 *Los Angeles Times* claims between 20 and 200 deaths occur each day in America's biggest cities because of dust and soot. The American Lung Association reported the cost of air pollution in the Los Angeles basin exceeded $14 billion in 1996, due to illness, property damage, crop losses, and corrosion.

Veziroglu completed another study in 1998 that involved an economic comparison of competing energy systems based on the effective costs of these fuels. Effective costs include utilization, efficiency, cost of fuel, and fuel consumption costs. External costs include environmental damage, governmental expenditures for pollution abatement, and military expenditures for protecting oil supplies. His study compared fossil fuels with solar hydrogen, assuming that one-third of the hydrogen could be produced from hydropower and/or wind power, and two-thirds from solar power. The effective cost of a fossil fuel system was 7.95 cents per kilowatt hour, for coal and a synthetic fuel system the cost was 9.91 cents, and for a solar hydrogen system it was 6.92 cents.

Veziroglu's study concluded that liquid hydrogen was the best transportation fuel when compared with liquid fuels such as gasoline, and gaseous hydrogen was the best gaseous transportation fuel. Hydrogen could be converted to useful energy forms through five processes at the user's end, including thermal, mechanical, and electric. In contrast, fossil fuels can be converted through only one process, flame combustion. Hydrogen was 39 percent more efficient than fossil fuels, and it was safer. Solar hydrogen energy could reverse global warming, and it has the lowest effective cost when including environmental factors. As gasoline shortages develop in the future and costs increase, the cost discrepancy between gasoline and hydrogen will magnify. In other words, cost should not hinder beginning the hydrogen transition.

To develop the hydrogen economy, realistic conclusions must be drawn. Developing a hydrogen infrastructure requires the present fossil fuel system. A hydrogen evolution will occur, not a revolution. Some technological breakthroughs remain necessary to achieve competitiveness in hydrogen, especially in fuel cell technology and high-density storage systems. Presently, all hydrogen projects need funding. Finally, political engineering to involve all relevant players is extremely important, including manufacturers, utilities, investors, government, the public, and environmental groups. A good start has been made.

TIMETABLE

Once the United States makes the decision to convert to a hydrogen economy, how long will it take? Should the transition be market driven or government influenced, or both? It will require a cooperative effort among government, auto manufacturers, oil companies, and the public. If more states enact California-like legislation requiring zero emissions, the transition will definitely accelerate. Some have suggested taxing all carbon dioxide-producing fuels. This would drive up the price at the fuel pump, another incentive for hydrogen conversion.

Much of Europe is already planning for the transition. The German government and a consortium of auto manufacturers and energy companies are planning a strategy for the gradual introduction of hydrogen as an alternative to fossil fuels. They estimate the transition period will last between ten and twenty years. During this period, internal combustion engine vehicles will be bivalent, meaning they will run on gasoline and hydrogen. BMW already has a bivalent prototype and has plans to commence commercial production of hydrogen engines. During this period, hydrogen PEM fuel cells will replace batteries on some prototype cars.

Various European countries and manufacturers have been developing hydrogen vehicles. They thought it might be wise to agree on standards regarding these vehicles. As a result, they established the European Integrated Hydrogen Project (EIHP) to help communicate and standardize aspects of the hydrogen economy. Within two years, they obtained unified regulations relating to registration of hydrogen-fueled vehicles. In May, 1998 the Transport Energy Strategy was established under the leadership of BMW and DaimlerChrysler. The goal is to identify the best fuel for the twenty-first century, after examining all the criteria, and subsequently to establish a strategy for its introduction to the market.

When testifying at the 1986 U.S. Senate hearings on hydrogen, T. Nejat Veziroglu stated we could establish a complete hydrogen energy system within fifty years. At the 2000 Forum on Converting to a Hydrogen Economy, Veziroglu said that by 2074, the transition should be complete.

Roy McAlister, president of the American Hydrogen Association, believes the transition to a hydrogen fuel economy in the United States could be made by 2015. The conversion to a hydrogen economy would create millions of high-quality jobs and provide full employment to the United States. With the full conversion to hydrogen, energy prices would decline, helping to curb inflation, asserts McAlister.

Maury Albertson, Ph.D., professor emeritus in civil engineering at Colorado State University, asserts a complete conversion to a hydrogen economy could take place between ten and twenty years. He believes a series of steps is necessary to make the transition. Because hydrogen is so expensive to produce from water electrolysis, Albertson says the production of hydrogen for the first several decades will have come from the existing petroleum industry. After twenty years, at a minimum, using renewable energy sources will become a reality.

For vehicles, the hydrogen transition will include gradually changing from gasoline and diesel to existing fuels such as propane, compressed natural gas, and ethanol. These fuels could be used separately or in combination with gasoline and diesel, and then in combination with hydrogen. Finally, hydrogen alone would become the ultimate fuel.

Albertson believes the 800 million existing vehicles (over 200 million in the United States) can be converted to use either fossil fuels alone, a combination of fossil fuels and alternative fuels, or use alternative fuel alone. New vehicles, produced in the next ten to twenty years can have a gradual transition from internal combustion engines that use gasoline to vehicles that use gasoline with alternative fuels, such as compressed natural gas (CNG) and/or hydrogen. Presently, CNG refueling stations exist in many cities around the United States. The existing infrastructure for CNG can support hydrogen vehicles, but of course must be expanded during the transition.

As fuel cells become more prevalent in new vehicles, Albertson believes they will gradually replace the internal combustion engine. The same infrastructure that has served hydrogen-converted vehicles can be used to refuel the new fuel cell vehicles.

At the end of twenty years, when the transition is complete, emissions from vehicles will be eliminated. During the transition period, the emissions will be reduced but not eliminated. Albertson believes stationary power stations could also make the transition to hydrogen within this twenty-year period.

Studies have shown that compressed hydrogen consumed in a fuel cell vehicle could cost less per mile than gasoline in internal combustion engines. This could occur if hydrogen was produced locally in small-scale electrolyzers or small-scale steam methane reformers. This low cost of hydrogen requires a paradigm shift in the method of producing hydrogen that would involve the production of many hydrogen fuel processors, much like home furnaces. These fuel processors would be produced when needed, as the market is penetrated. No massive pipeline system would be required, nor central hydrogen production plants, nor hydrogen liquefaction facilities, nor fleets of cryogenic tanker trucks.

If the industry built 1,000 fueling processors, each capable of supplying fuel for 1,000 fuel cell vehicles, the projected cost would be $230,000 per processor, or $230 per fuel cell vehicle. Compare this to the oil industry, which invests $11 billion a year in the United States to maintain the gasoline infrastructure, costing $1,000 per vehicle sold. One can conclude that the total fuel infrastructure costs for introducing and maintaining a hydrogen supply system might be less than the annual cost of maintaining the gasoline infrastructure.

During the transition, the dilemma of how to store hydrogen in the vehicle must be solved. We know that hydrogen is the ideal fuel for a fuel cell. Either 5,000 psi of compressed hydrogen or liquid hydrogen can be stored on a vehicle without compromising trunk space. Driving ranges for hydrogen cars will be reduced when compared with comparable internal combustion engines. Several auto manufacturers are using a metal hydride storage system.

Recent fuel cell technology by Ballard Power Systems may accelerate the transition process. Until recently, fuel cells were too large, cumbersome, and low on power to run a car. In the mid-1990s, Ballard Power Systems improved their fuel cell power to volume ratio. Up until that time, fuel cells delivered about 4.73 kilowatts per cubic foot. At this power production, the engine would occupy the entire trunk and backseat space. Five years ago, Ballard achieved 28.3 kilowatts per cubic feet. For the first time, fuel cells could fit under the hood. Ballard's latest fuel cell, the Mark 900, delivers 37.1 kilowatts per cubic foot, powerful enough to make a 75-kilowatt engine that fits comfortably under the hood without sacrificing cargo and passenger space. Ballard vice president Paul Lancaster promises that by 2004, a quarter million cells will be produced in Ballard's $400-million production plant each year. The company has contracted with Ford, General Motors, DaimlerChrysler, Nissan, and Honda to put the Mark 900 in their cars.

A fuel cell is used to power the NECAR 4, based on a Mercedes-Benz A-class compact sedan. The car accommodates five people plus luggage, reaches 90 miles per hour, and can go about 280 miles between fill ups.

Ferdinand Panik, director of DaimlerChysler's fuel cell project in Germany, believes hydrogen fuel cells will power a quarter of the new cars sold worldwide by 2020. Some believe this timetable may accelerate because oil companies can now support hydrogen. Don Huberts, chief executive officer of Shell Hydrogen in Amsterdam, predicts hydrogen will be the world's number one fuel in the twenty-first century.

Ballard's proton exchange membrane fuel cell converts fuel into power twice as efficiently as an internal combustion engine, with the benefits of no noise or noxious emissions. Gus McDonald, British Transport Minister, declared in December,1999 that fuel cells would allow more than half of new British cars to be "pollution free" within a decade. To be truly emission free, hydrogen must be produced from solar sources.

To make this transition to hydrogen, the public must be educated. Industry, politics, and society must accustom themselves to a completely new energy resource in their daily lives. Practical legislation and unified regulations must be passed to support this hydrogen transition. Public acceptance of hydrogen as a clean fuel will depend on campaigns by car and bus manufacturers when they introduce the technology. Government and industry must cooperatively explain the hydrogen story to encourage the consumer to trust and embrace hydrogen products. Only then can the transition be completed, allowing the planet Earth to reclaim its pristine heritage. Before we can proclaim hydrogen as the savior of our planet, it would be wise to examine other alternative fuels. Only then, can we make this proclamation.

Chapter Nine

ALTERNATIVE FUELS

Is Hydrogen the Best?

Dwindling fossil fuel reserves and deteriorating environmental conditions mean that we have but little choice to replace fossil fuels with alternative fuels. The ideal fuel should be sustainable, that is, not depletable. It should be renewable, capable of being replaced by natural ecological cycles or sound management practices. An ideal fuel would be environmentally friendly, producing no contribution to global warming or damage to the health of man, plants, or animals. Much research in the past several decades has given us viable alternatives to fossil fuels. Each alternative has its benefits, and each has its drawbacks. Many of these alternative fuels may play a role in our future energy economy as we wean ourselves from fossil fuels.

The term "alternative fuel" has been used to describe any fuel suggested for use in transportation vehicles other than gasoline or diesel fuel. The term also applies to any fuel that can replace fossil fuels for power generating plants. During the early 1900s, car buyers could choose from among internal combustion engines, steam engines, or electric vehicles. Henry Ford had even investigated the use of ethanol as a renewable home-grown fuel whose production benefitted agriculture. Both steam and electric vehicles made little noise and had good drive-ability, but both had a lengthy start-up time, steam engines were complex, and electric vehicles had a short driving range. With the introduction of electric starters, the internal combustion engine achieved a combination of economy, driving range, and ease of use that could not be matched.

The realization that oil-producing nations held the majority of the world's oil reserves and could dictate price and availability initiated consciousness for alternative fuels. The United States, Japan, and several European nations led the development of alternative fuels because they depended on imported oil to meet their transportation needs. Projected annual growth in light-duty vehicle fuel consumption is 0.8 percent, and for heavy-duty vehicles growth is projected at 1.3 percent through the year 2015. Growth in net oil imports is projected to be 1.9 percent, which suggests a widening gap between petroleum consumption and domestic production.

Environmental consciousness also developed about this time, and environmental advantages kept interest high for alternative fuels. Alternative fuel technology has also developed, making it feasible today to use these fuels in vehicles. The question now is, which alternative fuel can be produced at a competitive cost? Alternative fuel vehicles have become more prevalent throughout the United States because of the Energy Policy Act of 1992, known as EPACT. The act requires federal and state government, and companies producing alternative vehicles, to purchase these alternative fuel vehicles. By 1997, the federal government had acquired 15,000 alternative fuel vehicles. The U.S. Information Administration estimates that within ten years, annual sales could exceed 1 million vehicles, due to state mandates, EPACT, and market demand for alternative fuel vehicles. These vehicles will create a substantial demand for new fuel storage, transportation, and dispensing facilities, requiring modifications of existing facilities.

Alternative fuels for transportation include methanol, ethanol, natural gas, propane gas, vegetable oils, and hydrogen. After we examine these alternative fuels, one can only conclude that hydrogen is the best.

LIQUID FUELS FOR TRANSPORTATION

Methanol

Both methanol (methane alcohol) and ethanol are alternative transportation fuels. They are liquids and have several physical and combustion properties similar to gasoline and diesel fuel. These similar properties allow use of the same basic engines and fuel system technology as used with gasoline and diesel fuel. Both fuels have a higher octane rating than typical gasoline, which allows the alcohols to have a much higher compression ratio, resulting in increased thermal efficiency. A significant drawback is that they have a lower energy density, meaning fewer British thermal units (Btu) per gallon. In other words, it takes two gallons of methanol and one and one- half gallons of ethanol to take you as far as one gallon of gasoline. Current gasoline vehicles, retrofitted for alcohol fuel, do not allow for significantly increasing the fuel tank size.

Emissions from methanol and ethanol, compared with gasoline and diesel, contain less ozone. However, the disadvantage is that they produce formaldehyde and acetaldehyde as combustion products in larger quantities than the toxic compounds from petroleum fuels. The mass of emissions from methanol and ethanol are not significantly different from those of petroleum fuels. Either of these alcohols can be used in a diesel engine with the advantage of producing less soot, fewer particulates, and less nitrogen oxide.

Methanol, once referred to as wood alcohol, was originally made from the distillation of wood. Currently it is made from steam reformation of natural gas. Natural gas is reacted with steam in the presence of a catalyst under high heat to form carbon monoxide and hydrogen. These elements go through the methanol process to form methanol. Steam reformation has a thermal efficiency of about 56 to 62 percent.

Large plants can produce 2,000 to 2,500 metric tons of methanol per day. Plants as large as 10,000 metric tons have been proposed to serve transportation vehicles. In the United States, the primary methanol production sites are located on the Gulf Coast. Canada, South America, Europe, and the Middle East also produce methanol. Any country with remote natural gas reserves is a good candidate for methanol production. Methanol can also be produced from coal and municipal waste. Methanol is a colorless liquid used in industry as a solvent. In 1995, a total of 1.7 billion gallons of it was produced in the United States, ranking methanol the twenty-first most used chemical. The largest use occurred in the manufacture of methyl tertiary butyl ether

(MTBE), a gasoline additive. In 1995, 2.64 billion gallons of MTBE were produced, making it the twelfth most used chemical.

When the U.S. government mandated the phase-out of lead as a gasoline additive, low concentrations of methanol provided a good non-metallic lead substitute, used as a winter oxygenation additive. Methyl tertiary butyl ether (MTBE) is an octane blending agent made by reacting methanol with iso-butylene. Until recently, it was used in existing gasoline vehicles but has now been found to contaminate drinking water supplies. MTBE is currently being phased out as a gasoline additive. At its peak, it comprised 2.5 percent of the total gasoline supply.

Methanol's history as an alternative fuel dates back to the 1930s. It was used to replace or supplement gasoline supplies in high-performance engines in Grand Prix vehicles. For about two decades, it was the only fuel allowed in competing vehicles at the Indianapolis 500.

Flexible fuel vehicles can operate on gasoline, M85 (85 percent methanol and 15 percent gasoline), or any mixture of the two fuels. The engine computer control system automatically adjusts the air fuel ratio and the spark timing for the blend of methanol and gasoline. The system allows for the use of one fuel tank and can add about $100 to $400 to production costs.

Disadvantages of methanol include higher cost compared with gasoline, a reduced driving range, and a burning flame not visible in direct sunlight. Methanol's low vapor pressure does not allow it to support cold starts.

The advantage of methanol is that it is a convenient, familiar liquid fuel produced readily that uses proven technology. Vehicle manufacturers can easily design a car to use methanol that will outperform an equivalent gasoline vehicle and has reduced emissions and better efficiency. Methanol is a popular fuel for fuel cells that convert fuels into electricity.

A great advantage of methanol is that it does not produce soot or smoke. It is an attractive diesel fuel as it has no particulates. Nitrogen oxides are low because of methanol's latent vaporization heat. Because it contains oxygen, methanol facilitates leaner combustion, resulting in lower carbon monoxide emissions, and carbon dioxide emissions are about 94 percent of similar gasoline fuel vehicles. Unburned hydrocarbon is less reactive because it is primarily methanol. Methanol's capability to form ozone is about 50 percent less than that of gasoline. It contains no sulfur and makes only a minor contribution to acid rain. However, producing methanol from steam reformation via natural gas releases almost half the greenhouse gases that producing gasoline does. When looking at the entire fuel cycle, from resource through combustion, methanol has greenhouse gas emissions similar to those of gasoline.

Ethanol

Ethanol has long been considered a good spark-igniter engine fuel, and engines ran on ethanol very early in their development. Ethanol can be produced by the agriculture community, an idea developed in Brazil.

In the United States, ethanol fuel is produced almost exclusively with fermentation technology. The U.S. feedstock is primarily corn, although grains, potatoes, and beets can also be used. Starches are converted to sugars, which are then fermented. In Brazil, sugarcane is the preferred feedstock, whereas in France, grapes unsuitable for wine are used. Cost and capability of large-scale production governs the choice of feedstock as one needs to calculate the cost of producing the crop and of the fermentation process. The petroleum used to produce ethanol reduces the petroleum value of ethanol as an alternative fuel.

Ethanol can be used as a transportation fuel in three primary ways: (1) blended with gasoline, typically 10 percent, and known as gasohol; (2) as a compound of reformulated gasoline, directly or transformed into a compound called ethyl tertiary butyl ether (ETBE); or (3) used directly as a fuel with 15 percent gasoline known as E85. Almost all vehicles tolerate these fuels with advantageous emission benefits. Ethanol and ETBE blends in gasoline are approved by the EPA for mandated oxygenated fuel programs to lower carbon monoxide levels.

Currently the United States produces 1.1 billion gallons of ethanol per year and Archer Daniels Midland accounts for 700 million gallons a year. Almost all ethanol production plants are in the Midwest, near the cornfields. About 25 million gallons are imported from the Caribbean countries, which produce it from sugarcane.

The primary advantage of using an ethanol blend is that carbon monoxide emissions are reduced. However, ethanol has a slightly higher nitrogen-oxide emission than does gasoline. Some studies have found lower hydrocarbons and nitrogen oxides compared with gasoline, but tailpipe emissions were about the same. E85 produces acetaldehydes. The specific reactivity of ethanol exhaust emissions (gram ozone per gram emission) is 30 percent lower than gasoline.

Vehicle technology for E85 use is the same as for M85 use, allowing manufacturers to easily build E85 vehicles. When ethanol is blended with gasoline, few differences exist in drive-ability and performance.

In 1975, Brazil almost went bankrupt due to increased oil prices dictated by OPEC. The government decided to quickly substitute alcohol for gasoline in all vehicles. Through tax incentives and subsidies, Brazil created a large increase in sugarcane production to produce alcohol. Since 1979, no pure gasoline-powered light vehicle can be run in Brazil. Half of the automobiles are fueled by pure methanol, and the rest by gasohol, which is a mixture of 22 percent ethanol with gasoline.

The industry is no longer subsidized. Sugarcane ethanol is called bagasse, which supplies 100 percent of the heat for factories and 92 percent of the electricity. Air quality of Brazil has improved, and the bagasse/ethanol industry has created 700,000 new jobs.

Natural Gas

Natural gas supplies approximately 25 percent of the energy in the United States and is one of the world's most abundant energies. It is most commonly used for space heating, electricity generation, and in industry. Being a good spark-igniter for internal combustion engines, natural gas was used as a fuel in the early days of engine development.

The greatest drawback for its broad use in transportation has been the inability to store onboard in vehicles sufficient amounts of natural gas. Technological advances have been made that allow natural gas storage at high pressures (above 5,000 psi) in cylinders made of aluminum and carbon fiber. Compressed natural gas (CNG) is the preferred method for natural gas storage in vehicles, and liquified natural gas (LNG) storage is becoming more popular in heavy-duty vehicles. This allows heavy-duty vehicles the same operating range as with other liquid petroleum fuels. Storing natural gas as a liquid results in a storage system that is half the weight and volume of a compressed natural gas system. The cost and emission advantages of natural gas make it a popular alternative fuel.

The main constituent of natural gas is methane, the lightest and simplest hydrocarbon compound. Methane is formed from four sources: (1) organic matter decomposed in the presence of heat, (2) organic matter converted by microorganisms, (3) oil that produces methane in the presence of heat, and (4) coal that releases methane over time. Large reserves of natural gas are found at depths of 15,000 to 30,000 feet, called deep gas, depths where oil is not found. Deep gas is expensive to drill for, but the quantities are estimated to be very large.

Compared to gasoline, natural gas contains much less energy per volume. For this reason, transport over long distances is not practical in the liquified form. Liquified natural gas is created when natural gas is cooled to -260 degrees Fahrenheit, and it contains only 60 percent of the energy for the same volume of gasoline.

Because natural gas readily mixes with air, emissions of carbon monoxide are typically low. It also has a high-octane rating, making it a good spark-ignition fuel. Compressed natural gas vehicles have low exhaust emissions and no evaporative emissions. Because of the additional weight of storing natural gas for an equivalent amount of gasoline, vehicle acceleration is hindered, as is fuel economy.

Liquefied Petroleum Gas

Liquefied petroleum gas (LPG) includes several light hydrocarbons whose distinguishing characteristic is liquefying when put under modest pressure (less than 300 psi). Propane and butane are the most common LPGs. The LPG used in vehicles is essentially all propane, a fuel used in vehicles for the past sixty years. Propane represents the largest volume of alternative fuel in the United States, comprising 97 percent of alternative fuels in 1992. Vehicles using propane number between 220,000 and 370,000. Nearly all propane vehicles are converted gasoline vehicles, with auto manufacturers selling only a few propane vehicles.

Propane is stored under pressure to keep it in liquid form. Tanks that store propane are pressure containers, made from low-carbon steel, resulting in a modest price compared with CNG or LNG tanks. Propane is vaporized in a device called a converter, allowing the storage pressure of propane to vaporize it. The converter in vehicles uses engine coolants to warm the propane to ensure that it completely vaporizes when it passes on to the mixer. The mixer then mixes the propane and air in the proper ratio before it enters the engine.

Approximately half of liquid petroleum is produced in association with natural gas production, and the other half is produced during crude oil refining. It is undesirable to leave liquid petroleum gases in natural gas as they come out of the ground because of their tendency to liquefy under modest pressure.

The major use of propane is for home space heating and for commercial purposes. LPG is also an important feedstock for petrochemicals. In 1992, the total production of LPG in the United States was about 20 percent of the gasoline equivalent of road motor fuel. If all propane was used for vehicle fuel, it would represent 10 percent of the gasoline equivalent. The heating value of propane is 72 percent that of gasoline, taking 1.4 gallons of propane to equal 1.0 gallon of gasoline. As a liquid, propane weighs about 68 percent less than gasoline.

Similar to natural gas, propane vehicles have low exhaust evaporative emissions associated with the vehicle. The unburned hydrocarbons from propane are easier to oxidize in oxidation catalysts than methane, resulting in low unburned hydrocarbon emissions. Because propane is a gas, it mixes well with the air before entering the engine, resulting in low carbon monoxide emissions. Oxides of nitrogen emissions can be reduced to very low levels if the air-fuel ratios are kept at the proper value.

Most propane fuel systems used in light-duty vehicles have been the mechanical-control type, which meters propane in proportion to the amount of air used by the engine. Propane vehicles usually have poorer acceleration, poorer drive-ability, and poorer cold start performance than do gasoline systems. Few heavy-duty propane vehicles have been developed.

Vegetable Oils

Vegetable oils are considered a good alternative fuel for use in diesel engines. The most common type of crops for extracting vegetable oils include soybeans, sunflowers, peanuts, and Chinese tallow trees. Many other plants can also yield significant vegetable oil production per acre.

By reacting vegetable oils with methanol or ethanol, esters that improve the quality of fuel form. These esterfied versions of vegetable oils are called biodiesel. Manufacturers of biodiesel have been trying to get transit fleets to use this alternative fuel. A blend usually contains 20 percent biodiesel combined with diesel fuel, also known as B20. This blend produces favorable emission properties, resulting in reduced smoke, particulates, and gaseous emissions when used in typical transit buses. A significant decrease in hydrocarbons and carbon monoxide emissions occurs. Emission toxins are reduced, depending on the percentage of biodiesel used. Biodiesel has low sulfur content, another pollution advantage, and it contains no toxins and quickly biodegrades when spilled on the ground or in water.

Europe seems to be quite interested in the advantages of biodiesel as it may achieve 5 to 7 percent market share for diesel fuels in the European market. These estimates are based on current or planned construction of biodiesel production facilities. More than 1 million tons of biodiesel will be produced annually in the European Union. In Europe, chainsaw lubrication oils must be 80 percent biodegradable, providing another use for vegetable oils.

The major drawback at this time is the higher cost of biodiesel compared with other alternative fuels.

A biodiesel blend has approximately 13 percent less energy than typical diesel fuel. Engines adjusted for diesel fuel will experience a power loss of approximately 7 percent when using biodiesel. Engine oil dilution is a potential problem with biodiesel, as it is more prone to oxidation and polymerization than diesel fuel. This could cause thick sludge to develop, making the oil too thick to perform. Combustion residues of vegetable oils cause problems with fuel injection, piston rings, and oil stability.

ALTERNATIVE FUELS FOR POWER PLANTS

Nuclear Fission

Nuclear energy is an existing technology capable of supplying a significant amount of the energy now being supplied by fossil fuels and does not contribute to greenhouse gases. Today it accounts for nearly 20 percent of electric energy generation in the United States, and in France, it provides 75 percent.

The heaviest natural nucleus is uranium, consisting of 92 protons and 146 neutrons, called uranium 238. Breaking up a uranium nucleus is called fission, which releases a tremendous amount of energy. The neutron has been a favorite particle with which to bombard nuclei, because it is electrically uncharged. Once inside the nucleus, the neutron collides with nuclear constituents, shaking up the nucleus. The nucleus sheds excess energy when neutrons emit a particle or particles, or by emitting x-rays. The resultant nucleus is often unstable and will undergo additional changes, and such a nucleus is called radioactive. Half-life is the time it takes for half of the radioactive nuclei in a sample to decay. Half-lives can vary from billions of years to less than a second, and the half-life of radioactive uranium is 4.5 billion years.

When nuclear particles combine to form a new nucleus, the protons and neutrons give up some of their mass. This is converted to the binding energy that holds the nuclei together, according to Einstein's equation $E = mc^2$ (energy equals mass times the speed of light squared). The binding energy per proton of nuclei increases when going from the lightest to heaviest elements, up to iron. Beyond iron, the binding energy per proton becomes weaker until uranium is reached. When uranium undergoes fission, it gives up some of its mass, which is converted to energy. The energy released is estimated to be about 20 million times the energy released in a typical chemical energy reaction. Conversely, if hydrogen is converted to helium, a great deal of energy is liberated. This process, nuclear fusion, combines lighter nuclei to form heavier nuclei.

On average, 2.4 neutrons are released in each fission process, more than enough to continue the reactive process. Uranium 235, not uranium 238, is responsible for most of the fission.

Water absorbs the energy created from a nuclear reactor, and this energy is converted to electrical energy. The heated water creates steam, which turns the turbine to produce electricity.

Fission reactions split the uranium atom and produce tremendous amounts of energy. The process produces radioactive waste end products, which are extremely harmful. In the fission process, a variety of lighter elements are produced, such as barium, krypton, iodine, strontium, cesium, and many others. All of these end products are radioactive and emit electrically charged particles and gamma rays. Radioactive emissions can cause severe burns and all kinds of cancer in humans and animals. Some of the harmful waste, such as strontium, can enter the food chain, and when ingested, it replaces calcium in the bone.

Uranium is a nonrenewable resource, and the supply for nuclear plants will last about 100 years at current consumption rates. A breeder reactor creates plutonium and makes more energy available for use than it consumes. Breeder reactors increase the uranium fuel available for energy production by a factor of at least six. Plutonium creates a problem because it can be used directly to make a nuclear bomb. The United States has banned the construction of breeder reactors and has banned commercial plants from producing plutonium. While the United States wants to destroy all stocks of plutonium, Russia considers its stockpiles of plutonium an asset.

Public reaction to nuclear energy in the United States is very negative, and no more nuclear power plants are being built. Considered by many the safest nuclear reactor ever built, the Shoreham Reactor on Long Island has been mothballed and will be converted into a fossil fuel power plant. This reactor cost $2 billion to build.

Nuclear Fusion

For many years the public has been told that nuclear fusion will be the answer to our fossil fuel dilemma, but the technology remained years away. The road to this technology has been bumpy, and as we will see, the technology might not be feasible.

As we discussed in the first chapter, the Sun produces a vast amount of energy by combining four nuclei of hydrogen to form a helium nucleus, and the resulting two helium positrons have less mass than the four hydrogen nuclei. The excess mass available in this reaction is converted into energy ($E = mc^2$). This conversion is so great that one gram of hydrogen converted to helium yields the equivalent of 20 tons of coal. If we could duplicate this nuclear fusion reaction on Earth, our energy problems could be solved. We have an unlimited amount of hydrogen in water, and the end product of a fusion reaction is nonradioactive helium.

Fusion presents a major engineering problem, because hydrogen nuclei (the protons) are charged positively. These positively charged particles repel each other with a force that increases inversely as the square of their distance. In other words, the closer they come to each other, the more they repel each other. To overcome this repulsion, the hydrogen nuclei must approach each other at a high rate of speed. This can be accomplished by increasing the temperature of the proton to between 100 and 200 million degrees Fahrenheit. The Sun's temperature is around 15 million degrees, but it has the advantage of gravitational attraction that keeps a high density of hydrogen nuclei at high enough temperatures for fusion.

Scientists produced a fusion reaction on Earth when they exploded the hydrogen bomb. To achieve this desired temperature and high density for fusion, they set off the bomb by exploding a fission bomb inside it. This resulted in a high temperature and compression of protons, allowing fusion to take place.

Hydrogen, kept at a million-degree temperature, will have its electrons stripped from the protons, creating a gas-like condition called plasma. When the plasma touches the sides of a bottle or container, it cools down and cannot maintain a high enough temperature for fusion. This can be resolved by creating an external magnetic field in a direction parallel to the long axis of the container. The charged particles will spiral around in the direction of this magnetic field and be kept away from the walls. The two ends of the container are connected to form a torus. The fusion reaction using this configuration for its main chamber is called a "Tokamak."

The federal government funds fusion research, as no private company is willing to take this financial risk. For nearly 40 years, the United States, western Europe, Russia, Canada, and Japan have funded fusion research. Because of the high cost, countries have been forced to reduce their research investment. The Department of Energy closed down the Tokamak project at Princeton University. An international consortium, including Europe, Japan, and Russia, has developed among the interested countries that want to invest billions of dollars for fusion research. The project is called ITER, International Thermonuclear Experimental Reactor, and completion of this experimental project is projected to occur in 2008.

Scientists are not certain that no radioactive pollution will result from a successful fusion reaction. They are concerned that neutrons produced in generating fusion energy will react with the walls of the container to create radioactive products. It is almost certain that the components of the fusion device will become embrittled as a result of neutron bombardment.

Researchers are investigating a different approach for a successful fusion reaction, called inertial confinement, which tries to emulate the way the hydrogen bomb was made to explode. The U.S. Congress has appropriated $191 million to research this technology at an experimental National Ignition Facility. Critics question the potential of radioactivity if the process simulates a hydrogen bomb.

The future is not bright for nuclear fusion, as there are too many insurmountable problems that may be impossible to solve.

Water

Running water is an energy source supplied by the sun. The sun evaporates water from the lakes, rivers, and oceans, and the result is rain. Rain falling at higher elevations flows downhill, and can provide energy to generate electricity. In 1940, nearly 40 percent of the United States' electricity was generated by hydropower, and the original hydropower plants extracted the energy directly from running water. However, the one great disadvantage was that hydropower depended on the amount of water available, which was more in the spring and less in the fall. To remedy this fluctuation, dams were built to store water and to provide insurance against water shortages.

Hydropower in dams turns the turbines that in turn drive the generators that produce electricity. Today, hydropower supplies approximately 9 percent of the U.S. electricity, and 20 percent for the world. Construction costs to build hydroelectric dams are enormous, and public resistance has restricted dam building. Part of this resistance comes from environmentalists who do not want to damage the ecosystems.

The great advantages of hydropower include its nonpollution capabilities, water being a renewable resource, with no fuel cost. After the initial capital costs of dam construction, the energy is quite inexpensive. Reservoirs are good for flood control and a good source of irrigation water. The drawbacks are the damage to flora and fauna, and its effect on biodiversity. Although quite rare, dam breaks are a concern. In 1975, a dam broke in the Chinese province of Henan that killed 200,000 people and caused millions of others to die of famine and disease.

Wind Energy

Another energy dependent on the Sun is wind power. Because the Sun's energy is not absorbed by the Earth uniformly, some regions of the atmosphere are warmer than others. Warmer air expands and rises, while colder air constricts and settles. Expanded air is at a lower pressure than colder air, and the difference in pressure causes wind. Because of the Earth's spin, it drags some of the air close to it, and this causes the prevailing winds to blow from west to east.

When the United States was first settled, millions of windmills pumped water, and later generated electricity. Only after rural electrification became available did windmill sales plummet. Today, windmills are making a comeback and are a reliable and economical power source. By the twenty-first century, wind will generate 10 percent of Denmark's electricity. Denmark is one of the world's greatest suppliers of wind turbines. In California, wind power farms now provide the state with 1 percent of its electricity.

Wind turbines can respond to wind speeds of 9 to 60 miles per hour, and downtime on wind farms is between 2 and 5 percent, when insufficient wind blows. One windmill has a life span of 20 to 30 years. Wind turbines can now drive generators that produce 60-cycle alternating currents directly. Today's cost of electricity from wind is $0.05 per kilowatt-hour (kwh), down from $0.16 per kwh in 1985, and costs are expected to come down to $0.03 per kwh.

Research shows that taller towers built at higher elevations can better capture the winds where they are more consistent. Some estimate that wind power from North and South Dakota could supply 80 percent of the United States' electricity. Scientists believe that wind turbines by themselves could provide for 20 to 30 percent of the U.S. electrical needs when the expected fossil fuel crisis occurs. Offshore wind turbines could supply more than enough energy to satisfy worldwide demands.

Wind is a renewable, nonpolluting resource, and the energy is relatively cheap. Wind generators are modular and can be quickly added as demand increases. Wind power requires a large amount of space, and on some wind farms, machines must be placed 500 to 900 feet apart because of their great turbulence. An efficient wind farm must have at least 100 wind turbines. Because the turbines occupy only less than 5 percent of the space, the remaining space could be used for ranching and farming.

Disadvantages of wind farms include the noise created by the turbines and their visual unsightliness. Windy parts of the country are typically far removed from population centers where the energy is used. Electrical transmissions must rely on already established grids, at the mercy of utility companies. If the percentage of electricity supplied by wind is 15 percent or more, the grids would have to be redesigned. Some claim this figure should be 45 percent.

To have backup energy available, each wind farm could implement a battery system that allows storage of the energy when the wind is not blowing. As we will see, electricity generated by windmills may not have to be used directly. The electricity could be used for electrolysis, separating water into its constituents, hydrogen and oxygen. The hydrogen could then be used as a fuel.

Geothermal

Scientists estimate that nonsolar geothermal energy found in the Earth's interior is 35 billion times the amount of energy consumed annually. Approximately 60 percent was derived from the planet's formation and 40 percent from radioactive elements in the Earth's crust and mantle. Sadly, only a small percentage of this energy is available for use. If it could be tapped, geothermal sources could provide energy that would last thousands of years.

In Reykjavik, Iceland, 90 percent of space heating comes from geothermal energy. New Zealand and France have also tapped into their geothermal energy. Between 30 and 40 countries now use geothermal energy and are contemplating increasing their use of it. One-fourth of this energy provides heating and the remainder provides power generation. The European Community has projected an industrial capability of generating 20 megawatts of power by the year 2000. The Geysers, California, has one of the largest geothermal installations in the United States. The U.S. government has expressed a surprising lack of support for geothermal energy, considering its potential is as great as nuclear fusion energy. The cost of geothermal power plants compares favorably with fossil fuel plants, but experts do not foresee more than 1 percent of the world's energy coming from geothermal sources.

Geothermal reservoirs occur where hot liquid or gas escapes spontaneously from rock formation in the Earth's crust. Most of the energy in the Earth's crust is found in rocks heated by radioactivity or by primeval energy sources when the Earth was formed. Enough energy is found in these hot, dry rocks to supply all of the energy of the world for many centuries. In 1974, the Los Alamos National Laboratory received a patent for a mining method to tap the enormous amount of heat energy trapped in these rocks. They envisioned hot water vaporizing a hydrocarbon that would power an engine to drive an electric generator.

Ocean Tides

Tides provide an energy source not dependent on the Sun. In most regions of the Earth, all of the waters, including oceans, lakes, and rivers, rise at intervals of 12 hours and 25 minutes and then they recede. A great energy potential occurs when the water is at high tide. The rise and fall of the tides is attributed to the gravitational pull of the Moon.

Why are there two tides? The Earth and Moon rotate around their common center of gravity, which is deep within the Earth's interior. This rotation causes the water farthest away from the Moon to bulge, a centrifugal force effect. This causes a tide. The other tide is caused by the pulling of the water directly under the Moon. The Sun also plays a role, its effectiveness being 46 percent of that of the Moon. When the Earth, Moon, and Sun are in the same straight line, the tides are 46 percent higher than during quarter Moons.

Centuries ago, tidal mills were used in most of Europe. Incoming tide waters would enter a storage pond through a sluice, remaining until the tide turned back. The outgoing tides flowed back to the sea through a water well, providing energy. In 1960, a tidal installation was built to generate electricity in northern Brittany. A dam at the mouth of the bay, constructed with a sluice gate, allowed the water to enter the bay at high tide to form a lake. When the tide receded, the water permitted to leave the bay turned a turbine that operated a generator. The main advantage was its lack of environmental stress, and the disadvantage was its lunar dependency. It required a high tide of 16 feet to generate electrcity.

Ocean Thermal Energy Conversion (OTEC)

The amount of energy from the Sun that goes into the ocean is enormous. If only 0.01 percent were converted to useful purposes, this energy could supply the world's energy needs several times over. Water can efficiently absorb light energy from the Sun. Between the Tropic of Cancer (15 degrees north) and the Tropic of Capricorn (15 degrees south) the ocean is heated to about 82 degrees Fahrenheit. This temperature remains constant, more or less, day and night in all seasons to a depth of about 300 feet. Below 300 feet, the temperature steadily decreases, and at about 3,000 feet, it measures a constant 41 degrees Fahrenheit, no matter how deep. Cold water is heavier than warm water, and it does not mix with the water above.

This difference between the top and bottom layer has scientists believing it is possible to construct an engine to drive a turbine to generate electricity. The greater the temperature difference between the warm and cold layers, the greater the efficiency of the engine. During the 1970s, the U.S. government funded research to determine the feasability of OTEC as a competitive source of energy. In 1978, the State of Hawaii, the Lockheed Corporation, and Dillingham Corporation determined that it was possible to operate a sea installation to generate an amount of electricity greater than the amount to energize the equipment. Japan has a 20-megawatt plant operating on the island of Nauru. The pipe for the cold water needs to be about 40 feet in diameter, and the scale of the plant is equally enormous.

Professor John Bockris, of Texas A & M University, describes a method of heating a low-boiling point organic liquid by the surface water. The exit vapor could be condensed by the use of cold water pumped from the sea depth. A thermal cycle would occur by withdrawing heat from surface water and converting a smaller fraction to mechanical energy with the efficiency being 0.2 percent. The heat engines would sit on a floating platform and would drive conventional electric generators.

The advantages of OTEC are many: free fuel, no land cost, little maintenance costs for equipment; plants could operate without personnel for long periods, and unsightly plants would be out of view. Cables could transport electricity to land. Some have suggested shipping nitrogen from shore and converting hydrogen obtained by electrolysis into ammonia. However, the most intriguing use of OTEC energy is to make hydrogen and use the hydrogen directly as a source of energy. At the moment, the least expensive use is to generate hydrogen by electrolysis from the energy of OTEC. This in turn could be made into methanol that could be used directly in automobiles. The price of methanol would be comparable to the price of gasoline, with the drawback that burning methanol produces a greenhouse gas. Scientists also fear that OTEC would denude nutrients from the ocean depth.

The United Nations reports that 53 nations in the tropical regions might be candidates for OTEC technology.

Biomass

Biomass is vegetation and organic animal waste from recent times, as compared with fossil fuels from ancient times. Approximately 4 percent of the energy used in the United States comes from biomass, with half of that coming from firewood used in stoves for space heating. The remainder comes from corn and sugar used in ethanol. In Denmark, Austria, Sweden, and Finland, biomass meets about 10 percent of energy needs. In developing countries, biomass supplies an average of 36 percent of energy requirements. In India, the figure is 56 percent, and in Tanzania, it is 97 percent. The U.S. Department of Energy hopes that within twenty years, the United States could acquire 15 to 20 percent of its energy requirements from biomass. Likewise, the United Nations reported the world should derive 55 percent of its energy from biomass by the year 2050.

When biomass is grown, carbon dioxide is removed from the atmosphere and is returned when biomass is used as a fuel. Both the United Nations and U.S. government believe that biomass will not harm the environment. In contrast, fossil fuels removed carbon dioxide millions of years ago, but their use now returns carbon dioxide where it is doing great harm. Recently grown biomass delivers fewer toxic substances into the atmosphere than do fossil fuels. The least desirable way of providing biomass is to use old growth trees, and the most desirable sources are products from ongoing agriculture, such as sugar crops, and waste products, such as sawdust, scrap wood, and pulp waste. David Hall from the University College in London estimates that wastes could supply as much as 7.5 percent of the world's energy needs. We have discussed ethanol, a biomass used for fueling vehicle transportation. Corn is not as efficient a crop as sugarcane because corn stores its energy in the form of starch rather than sugar. Producing ethanol requires additional steps to convert starch to sugar. Futurists envision plantations of woody plants in the United States that could become a fuel economically competitive with gasoline.

As an important energy source, biomass plants must efficiently produce an energy-rich end product very quickly. Broadleaf plants that have a short growing season are efficient photosynthetically and are preferred. These plants should require little or no fertilizer and little water. Algae and kelp have been considered for biomass, and both grow in water. To be useful as a biomass, algae must first have the water removed. Most kelp reproduce their weight in four to six months. Both algae and kelp require that bacteria anaerobically digest the harvest to produce methane, which can be used directly as a transportation fuel. Algae production has occurred in Japan for many years, and California has experimental kelp farms.

Solar Energy

A renewable energy economy is based on solar energy, not on fossil fuels. Renewable energy is immune to the threat of future oil and gas shortages. According to a recent study by Shell International Petroleum Corporation, by 2050, renewable energy may provide almost as much energy as all fossil and nuclear energy combined. Solar energy can be collected through several pathways, including wind, the ocean thermal gradient, directly onto large area solar collectors, and by photovoltaic cells. Solar energy can be used to release hydrogen from hydrogen-containing compounds, such as water through electrolysis. Hydrogen then becomes a fuel, bypassing the fossil fuel cycle. This forms the basis of a solar-hydrogen economy.

Solar thermal techniques involve collecting solar radiation on a large area collector, focused onto a smaller area receiver, so the temperature is magnified greatly. The heat collected then drives an electric generator. Three principal types of solar thermal collectors exist: (1) parabolic trough collectors, (2) parabolic dishes, and (3) central-receiving collectors.

Parabolic trough collectors focus sunlight onto a central pipe that contains oil. The hot oil reaches a temperature of 300 to 400 degrees centigrade and passes through a heat exchanger that produces steam to drive a steam turbine generator. The overall efficiency of converting sunlight to electricity is about 13 to 17 percent. The majority of parabolic trough systems are in California, and have a capacity of 354 megawatts. They can produce energy for $0.12 to $0.16 per kilowatt hour, expected to drop to $0.08 to $0.12.

Parabolic dish systems consist of an array of parabolic dishes, with each tracking the sun in two axes. A small generator located at each dish converts the thermal energy to electricity or the receiver heats a working fluid that is then piped to a central location to produce steam and electricity.

Stirling engines have received the most attention for use with parabolic dishes. These engines use external heat to drive a working gas, usually hydrogen or helium, at high pressure to generate energy. The expansion and contraction of the gas moves a piston that is mechanically linked with a generator. Parabolic dishes are highly efficient because they have high concentration ratios and always point toward the sun, with current efficiencies between 16 and 24 percent. These systems are projected to produce electricity at costs of $0.06 to $0.12 per kilowatt hour.

Central-receiver systems consist of moveable flat-plate heliostats that focus sunlight onto a central-receiver tower that heats a working fluid. This drives a steam turbine to generate electricity. The Solar One site in Barstow, California generated 10 megawatts. Solar Two uses molten salt rather than water as the operating fluid. Electricity costs of $0.05 per kilowatt hour are expected.

Photovoltaics

Solar cells are tiny electric generators that seem to produce electricity out of nothing through a phenomenon called the "photovoltaic effect." They can provide power on a milliwatt scale for watches or for a million-watt power plant to supply electricity to a city. Because solar cells are light, reliable, and fuel-free, they have become standard power-generating equipment on all satellites. An entire industry developed to produce solar space cells, a cornerstone technolgy of the emerging solar industry.

Photovoltaic cells convert solar energy directly into electrical energy. In 1954, Bell Laboratories developed the first practical cell, and by the mid-1990s, the technology had achieved a 25 percent efficiency. It is infinitely renewable and nonpolluting and could satisfy almost all of the energy needs of the world.

All matter consists of a positive and negative electrical

charge. A photovoltaic cell is designed to have a potential difference in charge, and when light shines on the cell, it frees up electrons, which move because of their potential differences. The resulting electrical current can be drawn from the cell to do work. Because it has no moving parts, a photovoltaic cell will last indefinitely if protected from the environment. Photovoltaic cells can be connected to produce large amounts of energy.

The expense of generating electricity presents the biggest problem for photovoltaic cells. It costs $0.30 per kilowatt hour compared with $0.06 for fossil fuel energy, and this does not include the cost of environmental control and their harmful effects.

Photovoltaics have been used in remote relay stations, remote lighthouses, microwave relay stations, and road call boxes. The Japanese have sold more than 100 million products that use photovoltaics in hand calculators and wrist watches. Using photovoltaics to replace a kerosene lamp would cost about $500, too expensive for the 2 billion people who lack electricity in developing countries. However, funding from the World Bank and United Nations has enabled the installation of hundreds of thousands of photovoltaic devices on the roofs of homes in developing countries.

So how does it work? Silicon is a semiconductor used in photovoltaic technology. The first step to making a cell out of crystalline silicon involves adding an impurity, called doping. Arsenic, which has a single electron in its outermost shell, is added to the silicon, and its atoms take their place in the outermost crystal silicon structure. The outermost electrons of an arsenic atom will not contribute to the "glue" holding the silicon atoms, nor will it tightly bind to the atom. Under the action of a single force, it will move easily through the crystal to create a current. The combination of a silicon crystal with an arsenic impurity is called an "n" type semiconductor, with "n" standing for the negatively charged electron. Other impurities such as phosphorous or antimony will produce the same effect as does arsenic.

A "p" semiconductor involves gallium, added to crystalline silicon. Gallium's outermost shell is incomplete, and when added to silicon, gallium pulls an electron from a neighboring silicon atom and leaves a positively charged hole in the silicon atom. If a voltage was applied to the material, successive positive holes would be filled from successive silicon atoms. The progression of holes is equivalent to a positive charge, opposite to the motion of electrons. This is called a "p" semiconductor, a positive current.

A current can be brought outside the cell by externally connecting the "n" type conductor to the "p" type wire, resulting in work. This is initiated by adding the energy of shining light.

Why so expensive? Silicon is the element of choice for making a reliable photovoltaic device. It is the most plentiful element on earth but requires a great amount of effort to sufficiently purify it from its natural state. A photovoltaic cell requires silicon with no impurities, except for those added. The expensive purified silicon currently costs $40 per kilogram. To succeed in replacing fossil fuels, the cost must come down.

Some utility companies have begun using photovoltaic devices, such as Pacific Gas and Electricity in California, which has about 700 installations operated by photovoltaics. These are smaller installations, usually in remote areas, or they are used as a backup for regular operation systems. New England Electricity in Gardries, Massachusetts, installed photovoltaic panels on homes at a cost of $10,000 per panel. Each panel generates 2.2 kilowatts per hour, which is sufficient for the power needs of a typical home, including heating. Panels occupy 25 percent of the roof areas. Electricity costs and upkeep for the homeowners are negligible because the panels will last twenty to thirty years, and installation costs can be recovered in ten years.

An important obstacle is lack of sunlight. When sunlight is not available, some alternative or storage medium must be provided. The most promising is hydrogen, manufactured with the photovoltaic techniques.

The Department of Energy estimates that by the year 2010, photovoltaics will supply 1 percent of the energy in the United States. Many think this estimate is too conservative as some laboratory installations are producing energy by photovoltaics at a cost of $0.10 per kilowatt hour. In the eastern United States, fossil-fuel electrical power costs $0.12 per kilowatt hour. Photovoltaic cells can also power automobiles and be used as a substitute for large utility companies.

A solar-cell power plant could be built in less than a year. Solar cells are interconnected and laminated into durable, sealed units to form solar modules that deliver power in proportion to the number of component cells. They are easily sized to provide the required power. Modules can be easily and speedily installed, and a tracking device can keep them aimed at the sun. About 100,000 American homes already use photovoltaic power, mostly in rural areas.

The Sacramento Municipal Utility District has installed photovoltaic energy systems on hundreds of its customers' homes. Each system has a 30-year life that provides 263 megawatts of electricity over its life. This replaces the combustion of 27,970,000 cubic feet of natural gas and avoids the emission of 433,000 pounds of carbon dioxide. The company plans to install 100 units each year. The biggest market for photovoltaics is in developing countries.

CONCLUSION

As we have seen, each alternative fuel improves on the environmental dilemma caused by fossil fuels. However, most of these alternative fuels have their own problems, whether they contribute to global warming, produce radioactive wastes, have high cost, lack a technology, or depend on nature. As we have discovered, hydrogen can solve most of these problems. Hydrogen energy does not cause global warming, it is not radioactive, it is a renewable and sustainable resource, the technology is present, it can be economically competitive, it does not depend on nature, and hydrogen can be a storehouse of energy created by the Sun and wind. Hydrogen just may be the perfect fuel that saves our planet!

The hydrogen story provides a solution to global warming, dwindling fossil fuels, and the looming energy crisis. Once the story is known, it is only a matter of time before it starts playing itself out. How fast the hydrogen economy develops will depend on the public, and this is the reason for the book. What can the public do? Politicians listen to their constituents, and they too are looking for a solution to our energy crisis. They need to be educated. Zero-emission legislation in other states, similar to California, would accelerate the hydrogen economy. Industry will be marketing hydrogen automobiles in the near future. By purchasing these vehicles in mass, the public would encourage further development of this non-polluting technology. Other consumers could retrofit their present car to run on either hydrogen or gasoline. Some members of congress have recently suggested tax incentives (as much as $4000) to encourage consumers to purchase alternative-energy vehicles and energy-efficient homes. Additional incentives for businesses to invest in energy-efficient technologies have also been suggested.

On May 17, 2001, President George W. Bush announced his administration's new energy policy. His emphasis was on increased fossil fuel production and drilling, not conservation. The plan calls for nineteen hundred new electricity generating plants by 2020, costing more than $500 billion. These plants would be powered by fossil fuels and some by nuclear energy. If new plants are to be built, it would be logical to design them for hydrogen production from electrolysis through renewable resources like solar. His policy also proposes 38,000 miles of new gas pipelines, adding to the already 2 million miles of oil and gas pipelines. These pipelines should be designed to carry hydrogen as well as natural gas.

Bush's plan calls for expanded federal support for hybrid fuel cars, cars that run on both gasoline and electricity. A $4 billion tax credit is proposed to stimulate the sale of hybrid gas-electric vehicles and fuel cells. Also, $40 billion of federal spending and tax credits are proposed for research into renewable energy resources and fuel efficient heating and power equipment.

President Bush has the chance to establish his legacy as a champion of clean energy and a new economy by emphasizing hydrogen energy instead of fossil fuels. If this administration fails to grasp the opportunity now, it is inevitable that a future administration will put us onto the hydrogen path.

Perhaps the energy crisis of the new millennium has been a blessing in disguise. The crisis is making people aware that cheap fossil fuel energy is a thing of the past, and it is forcing the public and industry to examine alternatives. Once the hydrogen story is known, good leadership will guide us through the new millennium with optimism for both the environment and economy, because hydrogen energy will be our future economy!

REFERENCES

CHAPTER ONE: THE NEED FOR ALTERNATIVE ENERGY

Bechtold, Richard L., *Alternative Fuels Guidebook: Properties, Storage, Dispensing, and Vehicle Facility Modifications*, Society of Automotive Engineers, Inc., Warrendale, PA, 1997.

Berger, John J., *Charging Ahead: The Business of Renewable Energy and What it Means for America,* Henry Holt and Company, New York, 1997.

Borowitz, Sidney, *Farewell Fossil Fuels: Reviewing America's Energy Policy*, Plenum Trade, New York, 1999.

Brower, Michael, *Cool Energy: Renewable Solutions to Environmental Problems*, The MIT Press, Cambridge, MA, 1992.

Cashman, Ty, "The Hydrogen Vision," Proceedings - Forum on Converting to a Hydrogen Economy, Fort Collins, CO, 2000.

Corson, Walter H., *The Global Ecology Handbook: What You can do about the Environmental Crisis*, Beacon Press, Boston, 1990.

Lemonick, Michael, "Life in the Greenhouse," Time, April 9, 2001.

McAlister, Roy, "Facts that Every Citizen Should Know," Proceedings - Forum on Converting to a Hydrogen Economy, Fort Collins, CO, 2000.

McDonald, Joe, "Global Warming Worse than Believed," *Coloradoan*, Fort Collins, CO, January 23, 2001.

Merriam Webster's Dictionary, Tenth Edition, Merriam-Webster, Incorporated, Springfield, MA, 1993.

Meyers, Robert A., (ed.), *The Wiley Encyclopedia of Environmental Pollution and Cleanup*, Volume I and II, Wiley-Interscience, New York, 1999.

Norbeck, Joseph; Heffel, James; Durdin, Thomas; Tabbara, Bassam; Bowden, John; and Montano, Michelle, *Hydrogen Fuel for Surface Transportation*, Society of Automotive Engineers, Inc., Warrendale, PA, 1996.

Mwangi, George, "Warming May End Up Costing World $300B Every Year," Associated Press, *Coloradoan,* Fort Collins, CO, February 2001.

Query, Paul, "Balmy Alaska," Associated Press, *Daily Inter Lake*, Kalispell, Montana, March, 2001.

Recer, Paul, "Shrinking Ice Sheet Raises Warning Flag," Associated Press, *Coloradoan*, Fort Collins, CO, February 2001.

Revkin, Andrew C., "Mountain Retreat: Glacier's Erasure," *Rocky Mountain News*, Denver, February, 2001.

Shute, Nancy, "The Weather Turns Wild," U.S. News and World Report, February 5, 2001.

Steinberg, Meyer, "The Hydrogen Economy: Past, Present, and Future," Proceedings - Forum on Converting to a Hydrogen Economy, Fort Collins, CO, 2000.

Strohschen, Norbert, "Climate Change and Environmental Protection - A Global Interest for Insurers and Reinsurers," *On Energies of Change: The Hydrogen Solution*, Gerling Akademie Verlag, Munich, 2000.

The Hydrogen Future and Hydrogen Fuel Cell Commercialization Acts of 1995, Hearings - U.S. Senate, Subcommittee on Energy Research and Development, 1996.

The Hydrogen Research and Development Act, Hearings - U.S. House of Representatives, Committee on Science and Technology (105), March 12, 1986.

Veziroglu, T. Nejat, "Solar Hydrogen Energy System: A Permanent Answer to Energy and Environmental Problems," Proceedings - Forum on Converting to a Hydrogen Economy, Fort Collins, CO, 2000.

Vovelle, Christian, *Pollutants from Combustion Formation and Impact on Atmospheric Chemistry*, Kluver Academic Publishers, Dordrecht, 2000.

CHAPTER TWO: HYDROGEN

Borowitz, Sidney, *Farewell Fossil Fuels: Reviewing America's Energy Policy*, Plenum Trade, New York, 1999.

Britannica Junior Encyclopedia, Britannica Encyclopedia, Chicago, 1975.

Cox, Kenneth E., and Williamson Jr., K. D., *Hydrogen: Its Technology and Implications, Utilization of Hydrogen*, Volume IV, CRC Press, Inc., Boca Raton, FL, 1979.

Encarta Encyclopedia 2000, Microsoft Works, Suite 2000.

Lodhi, M. A. K., "Hydrogen Production from Non-Fossil Energy Sources, Renewable and Consumable," Proceedings - Forum on Converting to a Hydrogen Economy, Fort Collins, CO, 2000.

McAlister, Roy, "Facts that Every Citizen Should Know," Proceedings - Forum on Converting to a Hydrogen Economy, Fort Collins, CO, 2000.

Meyers, Robert A., (ed.), *The Wiley Encyclopedia of Environmental Pollution and Cleanup*, Volume I and II, Wiley-Interscience, New York, 1999.

Norbeck, Joseph; Heffel, James; Durbin, Thomas; Tabbara, Bassam; Bowden, John; and Montano, Michelle, *Hydrogen Fuel for Surface Transportation*, Automotive Engineers, Inc., Warrendale, PA, 1996.

Steinberg, Meyer, "The Hydrogen Economy: Past, Present, and Future," Proceedings - Forum on Converting to a Hydrogen Economy, Fort Collins, CO, 2000.

The Hydrogen Future and Hydrogen Fuel Cell Commercialization Acts of 1995, Hearings - Subcommittee on Energy Research and Development, U.S. Senate, 1996.

Veziroglu, Nejat T., "Solar Hydrogen Energy System: A Permanent Answer to Energy and Environmental Problems," Proceedings - Forum on Converting to a Hydrogen Economy, Fort Collins, CO, 2000.

CHAPTER THREE: THE HYDROGEN INFRASTRUCTURE

Albertson, Maury, "Hydrogen Transition," Proceedings - Forum on Converting to a Hydrogen Economy, Fort Collins, CO, 2000.

Bracha, Michael, "Hydrogen Production and Infrastructure," Proceedings - HYFORUM 2000, Munich, 2000.

Cox, Kenneth E., and Williamson Jr., K. D., *Hydrogen: Its Technology and Implications, Utilization of Hydrogen,* Volume IV, CRC Press, Inc., Boca Raton, FL, 1979.

Hipp, Eberhard, and Gruber, Christian, "Hydrogen Propulsion Systems for Transit Buses - Results of Operation in Passenger Transport," Proceedings - HYFORUM 2000, Munich, 2000.

Kluyskens, Dominique, "Clean Power Trade Around the World - Via Hydrogen," Proceedings - HYFORUM 2000, Munich, 2000.

Lemonick, Michael, "How to Prevent a Meltdown," Time, Spring, 2000.

Norbeck, Joseph; Heffel, James; Durbin, Thomas; Tabbara, Bassam; Bowden, John; and Montano, Michelle, *Hydrogen Fuel for Surface Transportation,* Society of Automotive Engineers, Inc., Warrendale, PA, 1996.

Steinberg, Meyer, "The Hydrogen Economy: Past, Present, and Future," Proceedings - Forum on Converting to a Hydrogen Economy, Fort Collins, CO, 2000.

The Hydrogen Research and Development Act, Hearings - U.S. House of Representatives, Committee on Science and Technology (105), March 12, 1985.

Timm, Manfred, "Hydrogen and the Electric Utility Industry," *On Energies-of-Change: The Hydrogen Solution,* Gerling Akademie Verlag, Munich, 2000.

Wolf, Joachim, "Liquid Hydrogen Technology for Vehicles," Proceedings - HYFORUM 2000, Munich, 2000.

CHAPTER FOUR: FUEL CELLS

Anderson, Byron, "How to Tank and Tube for Renewable Fuels," Proceedings - Forum on Converting to a Hydrogen Economy, Fort Collins, 2000,

Bechtold, Richard, L., *Alternative Fuels Guidebook: Properties, Storage, Dispensing, and Vehicle Facility Modification*, Society of Automotive Engineers, Inc., Warrendale, PA, 1997.

Borowitz, Sidney, *Farewell Fossil Fuels: Reviewing America's Energy Policy*, Plenum Trade, New York, 1999.

Kesten, Martin, "Comparison of Hydrogen Storage Systems," Proceedings - HYFORUM 2000, Munich, 2000.

Kruger, Paul, "Electric Power Requirements in California for Large Scale Production of Hydrogen Fuel," Proceedings - Forum on Converting to a Hydrogen Economy, Fort Collins, CO, 2000.

Norbeck, Joseph; Heffel, James; Durbin, John; Tabbara, Bassam; Bowden, John; and Montano, Michelle, *Hydrogen Fuel for Surface Transportation*, Society of Automotive Engineers, Inc., Warrendale, PA, 1996.

The Hydrogen Research and Development Act, Hearings - U.S. House of Representatives, Committee on Science and Technology (105), March 12, 1986.

Timm, Manfred, "Hydrogen and the Electric Utility Industry," Proceedings - HYFORUM 2000, Munich, 2000.

Wakefield, Ernest, *History of the Electric Automobile: Hybrid Electric Vehicles*, Society of Automotive Engineers, Inc., Warrendale, PA, 1998.

Williams, Lawrence, "Energy in the Twenty-first Century," Proceedings - Forum on Converting to a Hydrogen Economy, Fort Collins, CO, 2000.

CHAPTER FIVE: HYDROGEN SAFETY

Cox, Kenneth E., and Williamson Jr., K. D., *Hydrogen: Its Technology and Implications, Utilizations of Hydrogen*, Volume IV, CRC Press, Inc., Boca Raton, FL, 1979.

Norbeck, Joseph; Heffel, James; Durbin, Thomas; Tabbara, Bassam; Bowden, John; and Montano, Michelle, *Hydrogen Fuel for Surface Transportation*, Society of Automotive Engineers, Inc., Warrendale, PA, 1996.

McAlister, Roy, "What Caused the Hindenburg Disaster?," Proceedings - Forum on Converting to a Hydrogen Economy, Fort Collins, CO, 2000.

CHAPTER SIX: PRACTICAL APPLICATIONS FOR HYDROGEN ENERGY

Cox, Kenneth E., and Williamson Jr., K. D., *Hydrogen: Its Technology and Implications, Utilization of Hydrogen*, Volume IV, CRC Press, Inc., Boca Raton, FL, 1979.

Esselstyn, Erik C., "One Cutting Edge of the Hydrogen Revolution: A Look at Four Demonstration Projects," Proceedings - Forum on Converting to a Hydrogen Economy, Fort Collins, CO, 2000.

Kluyskens, Dominique, "Clean Power Around the World-via Hydrogen," Proceedings - HYFORUM 2000, Munich 2000.

Meyer, Olaf, "Hydrogen Powered - A Brand New Car?," Proceedings - HYFORUM 2000, Munich 2000.

Norbeck, Joseph; Heffel, James; Durbin, Thomas; Tabbara, Bassam; Bowden, John; and Montano, Michelle, *Hydrogen Fuel for Surface Transportation*, Society of Automotive Engineers, Inc., Warrendale, PA, 1996.

Sapru, K., Tan, Z., and Ramachandran, S., "Potential for Early Commercialization of Renewable Hydrogen Technologies," Proceedings - Forum on Converting to a Hydrogen Economy, Fort Collins, CO, 2000.

Winter, Carl Jochen, *On Energies of Change: The Hydrogen Solution*, Gerling, Munich, 2000.

CHAPTER SEVEN: HYDROGEN'S CURRENT STATUS

Arnason, Bragi, and Sigfusson, Thorsteinn I., "Iceland - A Future Hydrogen Economy," Proceedings - Forum on Converting to a Hydrogen Economy, Fort Collins, CO, 2000.

Bracha, Michael, "Hydrogen Production and Infrastructure," Proceedings, HYFORUM 2000, Munich, 2000.

Esselstyn, Erik C., "One Cutting Edge of the Hydrogen Revolution: A Look at Four U.S. Demonstration Projects," Proceedings - Forum on Converting to a Hydrogen Economy, Fort Collins, CO, 2000.

Ganke, Toshihko: Kawasaki, Kensuke; Miki, Hirofumi; and Sugimoto, Kenji, "Renewable Power Sources," Proceedings - HYFORUM 2000, Munich, 2000.

Lloyd, Alan, "Hydrogen and Clean Air Regulations," Proceedings - HYFORUM 2000, Munich, 2000.

Teltschik, Horst, "Sustainable Business - The BMW Group Takes Its Responsibility for the Future," Proceedings - HYFORUM 2000, Munich, 2000.

Zweig, Robert M., and Provenzano, James J., "Pollution Solution Evolution, Cancer Prevention Option: New Opportunities for Hydrogen in California," Proceedings - Forum on Converting to a Hydrogen Economy, Fort Collins, CO, 2000.

CHAPTER EIGHT: MAKING THE TRANSITION TO HYDROGEN

Albertson, Maury, "Hydrogen Transition," Proceedings - Forum on Converting to a Hydrogen Economy, Fort Collins, CO, 2000.

Anderson, Byron, "How to Tank and Tube for Renewable Fuels," Proceedings - Forum on Converting to a Hydrogen Economy, Fort Collins, CO, 2000.

Bracha, Michael, "Hydrogen Production and Infrastructure," Proceedings - HYFORUM 2000, Munich, 2000.

Cox Kenneth E., and Williamson Jr., K. D., *Hydrogen: Its Technology and Implications, Utilization of Hydrogen,* Volume IV, CRC Press, Inc., Boca Raton, FL, 1979.

Teltschik, Horst, "Sustainable Business - The BMW Group Takes Its Responsibility for the Future," Proceedings - HYFORUM 2000, Munich, 2000.

Kluger, Jeffrey, "A Time to Despair," Time, April 9,2001.

Kluyskens, Dominique, "Clean Power Trade Around the World - Via Hydrogen," Proceedings - HYFORUM 2000, Munich, 2000.

Lemonick, Michael, "How to Prevent a Meltdown," Time, Special Edition, Spring 2000.

Lodhi, M. A. K., "Hydrogen Production from Non-Fossil Energy Sources, Renewable and Consumable," Proceedings - Forum on Converting to a Hydrogen Economy, Fort Collins, CO, 2000.

McAlister, Roy, "Facts That Every Citizen Should Know," Proceedings - Forum on Converting to a Hydrogen Economy, Fort Collins, CO, 2000.

Pearce, Fred, "Kicking the Habit," New Scientist, London, November 2000.

The Hydrogen Research and Development Act, Hearings - U.S. House of Representatives, Committee on Science and Technology (105), March 12, 1986.

The Hydrogen Future and Hydrogen Fuel Cell Commercialization Acts of 1995, Hearings - U.S. Senate, Subcommittee on Research and Development, 1996.

Thomas, C. E., "On Future Fuels: A Comparison of Options," Proceedings - HYFORUM 2000, Munich, 2000.

Timm, Manfred, "Hydrogen and the Electricity Utility Industry," Proceedings - HYFORUM 2000, Munich, 2000.

Wolf, Joachim, "Liquid Hydrogen for Vehicles," Proceedings - HYFORUM 2000, Munich, 2000.

CHAPTER NINE: ALTERNATIVE FUELS

Bechtold, Richard L., *Alternative Fuels Guidebook: Properties, Storage, Dispensing, and Vehicle Facility Modifications,* Society of Automotive Engineers, Inc., Warrendale, PA, 1997.

Berger, John J., *Charging Ahead: The Business of Renewable Energy and What It Means for America,* Henry Holt and Company, New York, 1997.

Bockris, J. O. M., *Energy: The Solar-Hydrogen Alternative,* John Wiley and Sons, New York, 1975.

Borowitz, Sidney, *Farewell Fossil Fuels: Reviewing America's Energy Policy,* Plenum Trade, New York, 1996.

Espo, David, "Democrats Unveil Energy Plan, Coloradoan (Associated Press), Fort Collins, CO, May 16, 2001.

Fouier, Ron, "Bush Sees 'Darker Future' Without Plan," Denver Rocky Mountain News, May 18, 2001.

Merriam Webster's Dictionary, Tenth Edition, Merriam-Webster, Incorporated, Springfield, MA, 1993.

Meyers, Robert A., (ed.), *The Wiley Encyclopedia of Environmental Pollution and Cleanup,* Volume I and II, Wiley-Interscience, New York, 1999.

Norbeck, Joseph; Heffel, James; Durbin, Thomas; Tabbara, Bassam; Bowden, John; and Montano, Michelle, *Hydrogen Fuel for Surface Transportation,* Society of Automotive Engineers, Inc., Warrendale, PA, 1996.

INDEX